Electronics for Techr

# Electronics for Technicians

**B. F. R. Gillman**
C. Eng., M.I.E.R.E.

*Lecturer in Electronics,
Southgate Technical College*

**B. A. Hudgell**
B.Sc., C.Eng., M.I.E.E.

*Senior Lecturer in Electronics,
Southgate Technical College*

HODDER AND STOUGHTON
LONDON   SYDNEY   AUCKLAND   TORONTO

**British Library Cataloguing in Publication Data**
Gillman, B
  Electronics for technicians, Level 2.
  1. Electronic apparatus and appliances
  I. Title  II. Hudgell, B
  621.381    TK7870

  ISBN 0–340–23441–5


First published 1979

Copyright © B. Gillman and B. Hudgell

All rights reserved. No part of this publication may be reproduced
or transmitted in any form or by any means, electronic or mechanical,
including photocopy, recording, or any information storage and retrieval
system, without permission in writing from the publisher.

Printed and bound in England
for Hodder and Stoughton Educational,
a division of Hodder and Stoughton Limited,
Mill Road, Dunton Green, Sevenoaks, Kent by
Richard Clay (The Chaucer Press), Ltd., Bungay, Suffolk

# Preface

The topics dealt with in this book illustrate a variety of fundamental principles and techniques which will be of value to anyone studying electronics, telecommunications or electrical engineering. The depth of treatment is aimed at Level II of the Technical Education Council Programmes, and the main objectives of the TEC Electronics II U76/010 Unit have been covered. The authors have not confined themselves to the content of the Unit and, where it has been considered worthwhile, certain sections have been expanded. This has resulted in each topic area being more complete, and reaching useful conclusions.

Although the book has primarily been written for technicians, it should form a useful introductory work for students following more advanced courses. Also, it is felt that the style and presentation of the material will make it particularly valuable to persons who are studying electronics privately and without the help of formal teaching.

The authors would like to express their gratitude to their wives for the patience and understanding they showed during the many hours spent preparing this book.

B.F.R.G.<br>
B.A.H.

# Contents

| | | |
|---|---|---|
| Preface | page | v |
| 1 – Semiconductor Materials | | 1 |
| 2 – The *p–n* Junction and its Characteristics | | 11 |
| 3 – Rectification and Simple Power Supply Circuits | | 22 |
| 4 – The Bipolar Transistor | | 39 |
| 5 – The Transistor Amplifier | | 59 |
| 6 – *L-C* Oscillators | | 81 |
| 7 – Waveform Generators | | 94 |
| 8 – Thermionic Valves | | 114 |
| 9 – The Cathode Ray Tube | | 130 |
| 10 – Logic Gates and Circuits | | 145 |
| Answers to Problems | | 159 |
| Index | | 163 |

# Chapter 1
# Semiconductor Materials

**1.1 Introduction**
At its most fundamental level, any study of electrical engineering is concerned with the pattern of movement of *electrons*. When electrons are made to move in a predictable manner they can produce useful effects. Some obvious examples of these effects are to light a lamp, turn a motor or produce sound from a loudspeaker. Why some materials should freely permit these electron movements, whereas others hardly allow them at all, requires an examination of the properties of the *atoms* of which they are formed.

**1.2 Electrons and their energy levels**
An accepted idea of the atom suggests it to be made up from a positively charged *nucleus* with orbiting, negatively charged electrons. The number of electrons is such that their total negative charge corresponds to the positive charge held by the nucleus. The atom therefore appears electrically neutral.

The radius of orbit of any particular electron is related to its energy level, the radius increasing as the electron's energy is increased. However, for a single isolated atom, only certain discrete electron energy levels are possible. The discrete energy values fall into groups called *shells*. Electrons in the inner shells are tightly bound to the

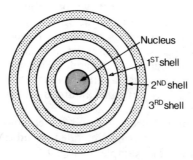

Fig 1.1 Diagrammatic representation of an atom showing the grouping of the possible electron orbits into shells.

nucleus and have low energies. Electrons in outer shells have higher energies and are more easily released from the atom. The electrons occupying the outermost shell are called the *valence* electrons and, of course, these are the most loosely bound. Also, these are the most important electrons as far as the electrical properties of a material are concerned.

It is possible for an electron to move to a higher unoccupied energy level if it acquires the necessary increase in energy. This energy will be provided from external sources such as heat or light.

### 1.3 Energy bands

In solids many atoms exist in close proximity with considerable interaction between them, the electrons being influenced not only by their own nucleus, but also by the nuclei of neighbouring atoms. The result of this is to increase the range of possible energy levels which the electrons can occupy. This effect is most marked for the valence electrons, for which a large number of closely spaced energy levels become available. The closeness of the spacing is such that it becomes convenient to consider that a continuous band of permissible energy levels, known as the *valence band*, exists.

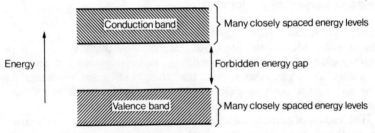

Fig 1.2 Representation of continuous energy bands in a solid.

Energy bands can be represented in a simple energy band diagram such as the one shown in Fig. 1.2 which is representative of a certain class of solids. There are in fact two permitted energy bands shown in this diagram. The lower one is the valence band which has just been discussed. The higher one is called the *conduction band* which is a further range of permitted energy levels. These two energy bands are separated by a forbidden energy gap, i.e. a range of energies denied to electrons. This means that a valence electron can be elevated to the conduction band only if it gains an amount of energy at least equal to the energy gap.

In Fig. 1.2 the valence band and the conduction band are shown shaded for clarity. This is not intended to indicate whether a band is full or otherwise. For example, in Fig. 1.2, if the energy gap is large, then at low temperatures the valence band will be full and the conduction band empty. At higher temperatures, some electrons may gain

# Semiconductor Materials

sufficient energy to reach the conduction band which is then partly filled. This of course leaves a corresponding number of unoccupied energy levels in the valence band.

These continuous energy bands are important because they enable the movement (or *drift* as it is known) of charge to take place when an external electric force is applied to the material.

Consider Fig. 1.2 once more. At low temperatures the valence band is full and, since the electrons do not have sufficient energy to reach the conduction band, there are no vacant energy levels to which they are able to move. A potential difference applied across the material cannot therefore cause an electric current. At higher temperatures however, a number of electrons may gain sufficient energy to reach the conduction band (this is a continuous process in which electrons are elevated to the conduction band when they gain sufficient energy and, a short time later, fall back to the valence band when they give up that energy). On balance though, we can consider this to be equivalent to having a number of electrons raised from the valence band to the conduction band on a more permanent basis. Once in the conduction band, a continuous range of permitted energy levels is available to them. This means that a small energy change can cause the electrons to change to new levels within the band and, in so doing, change their positions within the material. Such electron movements are generally random in nature and so there is no net charge movement in any particular direction. However, when an external potential difference is applied, the electric field will cause a net drift of electrons towards the more positive side. That is, electrical conduction occurs within the material. The rate of drift of charge is of course the electric current.

## 1.4 Conductors, insulators and semiconductors

Figure 1.3 shows the relative positions of the valence and conduction bands for insulators, conductors and semiconductors.

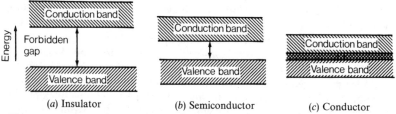

(a) Insulator     (b) Semiconductor     (c) Conductor

Fig 1.3 Relative spacing of valence and conduction bands for an insulator, semiconductor and conductor.

For materials classed as insulators, Fig. 1.3(a) shows that the energy difference between the valence and conduction bands is comparatively large. Consequently, at normal ambient temperatures, the number of valence electrons that acquire sufficient energy to jump to the conduction band is insignificant. The electrical resistivity of insulators is therefore extremely high.

From Fig. 1.3(c) it can be seen that for materials classed as conductors, the valence band actually overlaps the conduction band, and so there are electrons in the conduction band at all times. Here then is a situation of electrons in an energy band, with a continuous range of vacant energy levels available to them. The electrical resistivity of conductors is therefore extremely low.

Figure 1.3(b) represents a group of materials for which the energy gap between the valence and conduction bands is small compared with the gap for an insulator. At normal ambient temperatures a reasonable number of electrons gain sufficient thermal energy to jump into the conduction band. The number of electrons in the conduction band is small compared with conductors but is large when compared with insulators. The resistivity of such materials therefore lies somewhere between that of conductors and insulators and they are, quite aptly, described as *semiconductors*. Examples of semiconductor materials are copper oxide, selenium, gallium arsenide, germanium and silicon.

## 1.5 Holes

When an electron in a semiconductor jumps to the conduction band, it leaves behind a vacant space (i.e. a vacant energy level) in the valence band. This vacant space, or *hole* as it is known, can be filled by another valence electron from another nearby atom in the material. If this occurs the original hole will disappear but a hole will now exist in the position from which this electron came. Since energy levels in the valence band are continuous, this movement of holes will be caused by any small energy change. Hence, when a potential difference is applied to the material, a continuous movement of charge can occur. An imperfect, but useful representation of this hole movement is shown diagrammatically in Fig. 1.4.

The circle represents a hole created as a result of an electron being raised to the conduction band. The black dots represent electrons with valence band energies. Notice how the hole progresses through the material towards the negative side of the applied p.d. Its progression is of course due to electron movements in the opposite direction. This representation is rather crude because energy changes are implicit in this action, and electron movements are of a more random nature than the simple regular movement suggested. However it is convenient to consider the holes as *positive charge carriers* which are able to drift from the positive side to the negative side of the applied p.d. This charge movement in the valence band contributes to the total semiconductor current.

## 1.6 Intrinsic conduction

When an electron in a semiconductor acquires sufficient thermal energy to jump to the conduction band, a hole appears in the valence

# Semiconductor Materials

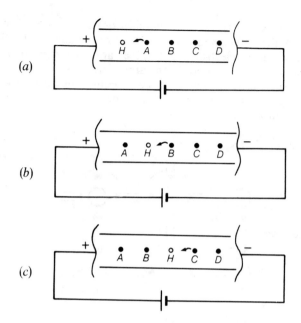

Fig 1.4 Representing the drift of a hole in a semiconductor material.
(a) 'H' represents a hole created as a result of an electron being elevated to the conduction band
(b) Electron 'A' fills original hole. A hole now occupies the position originally held by 'A'
(c) Electron 'B' fills hole left by movement of 'A'. Hole 'H' now occupies position originally held by 'B'

band. Electrons and holes are obviously generated in pairs. The average time for which any particular electron remains in the conduction band is very short, since it will quickly fall back into the valence band and fill a hole. This process is known as *recombination*. For any given temperature, the rate of production of electron-hole pairs tends to equal the rate of recombination, and so the average number of available charge carriers remains constant.

The total current in the semiconductor is the sum of the currents due to the electrons in the conduction band, and the holes in the valence band. Since, for a given applied p.d., conduction will depend on the number of electrons and holes available, the magnitude of the current will be temperature dependent. This temperature dependent conduction in a semiconductor is called *intrinsic* conduction.

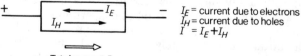

Fig 1.5 Components of current in a semiconductor.

### 1.7 Covalent bonds

The atoms of the semiconductor materials germanium and silicon, each have four valence electrons, that is, they are *tetravalent*. In a pure sample of either of these materials, the atoms arrange themselves in a regular pattern or *lattice*, each atom being equidistant from its four immediate neighbours. A two-dimensional impression of this arrangement is shown in Fig. 1.6.

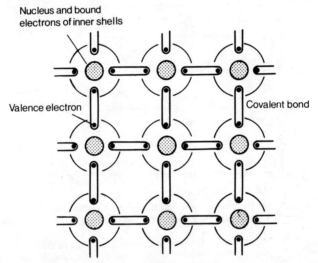

Fig 1.6 Two-dimensional impression of the regular crystal structure of a sample of pure silicon (or germanium), i.e. Intrinsic semiconductor.

It is found that each valence electron is shared by two adjacent atoms forming what is called a covalent bond between them. The effect of the covalent bonding is to bind the adjacent atoms firmly together and consequently hold the valence electrons more firmly in their locations. When a covalent bond is broken an electron moves away leaving a hole and so an electron-hole pair is generated.

### 1.8 N-type semiconductor

*N*-type is the term used to describe a semiconductor in which electrons predominate over holes as the charge carriers. The chemical elements antimony, arsenic and phosphorus each have five valence electrons per atom, that is, they are *pentavalent*. It is possible to introduce a minute amount (typically about one part in $10^7$ to $10^8$) of one of these elements into a sample of germanium or silicon, without significantly disturbing the regular atomic structure outlined in section 1.7. Because the amount introduced is extremely small, it is likely that the four immediate neighbours of any of the impurity atoms will be atoms of the pure semiconductor. These semiconductor atoms, in trying to re-

# Semiconductor Materials

gularise the atomic structure, will set up covalent bonds with the impurity atom. However, since only four valence electrons are required from any single atom to complete the lattice, the fifth electron will be left unbonded. This situation is shown diagrammatically in Fig. 1.7.

The fifth valence electron of the impurity atom, will not be as tightly bound into the atomic structure as those that have formed covalent bonds, and the amount of energy required to raise it to the conduction band is therefore relatively small.

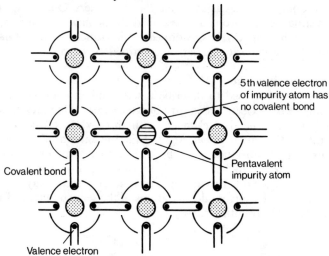

Fig 1.7 Two-dimensional impression of a pentavalent impurity atom surrounded by the tetravalent atoms of silicon (or germanium), i.e. *n*-type semiconductor material.

The diagram of Fig. 1.8 illustrates the relative energy level of these unbonded electrons. The fact that this lies in what for the semiconductor material is a forbidden energy region is due to it being part of a different material with different energy distributions. Also, this energy

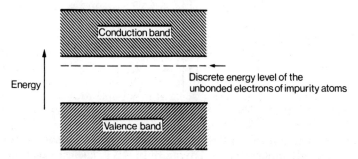

Fig. 1.8 Energy band diagram showing discrete level of the unbonded electrons of impurity atoms.

position is a discrete level rather than a band since the spacing between the impurity atoms is, on an atomic scale, very large.

The level of thermal energy available to the atomic structure in the normal range of ambient temperatures is such that most of these unbonded valence electrons are in the conduction band. The 'doping' of the pure semiconductor material therefore results in a substantial increase in the number of electrons in the conduction band. Electron-hole pairs will be thermally generated as before but electrons will now always outnumber holes as available charge carriers. Quite reasonably for this situation, electrons are described as the *majority* carriers, and holes as the *minority* carriers. A doped semiconductor material having electrons as the majority carriers is described as *n*-type.

## 1.9 P-type semiconductor

*P*-type is the term used to describe a semiconductor in which holes predominate over electrons as the charge carriers. Aluminium, boron and gallium are chemical elements with three valence electrons per atom, that is, they are *trivalent*. Again, a minute amount of one of these elements can be introduced into pure silicon or germanium without significant alteration of the regular pattern of the atomic structure. As before, because the quantity introduced is so very small, any impurity atom is likely to be surrounded by atoms of the semiconductor. Figure 1.9 illustrates the resulting situation.

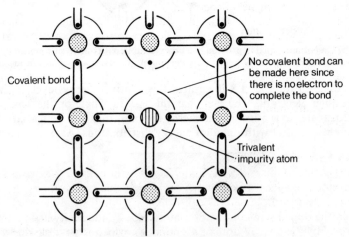

Fig 1.9 Two-dimensional impression of a trivalent atom surrounded by the tetravalent atoms of silicon (or germanium), i.e. *p*-type semiconductor material.

The diagram shows that the covalent bond structure is incomplete due to the impurity atom having only three valence electrons. If an electron becomes available, the impurity atom will acquire it in order to complete the unmade bond. Now it is found that a small amount of

## Semiconductor Materials

energy acquired by any nearby valence electron, will enable it to break its covalent bond and fill this gap. The amount of energy required is far less than that needed for the same electron to jump to the conduction band. In effect, the impurity atom has provided an energy level at which electrons can exist, in between the valence and conduction bands. This situation is illustrated in Fig. 1.10.

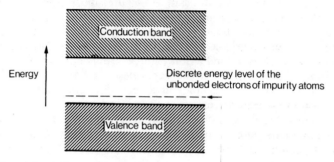

Fig. 1.10 Energy band diagram showing discrete energy level introduced by trivalent impurity atoms.

The level of thermal energy available in the normal range of ambient temperatures, ensures that most of these intermediate energy positions introduced by the trivalent impurity atoms are filled. The result, of course, is holes in the valence band.

From the above, it can be seen that when germanium or silicon is doped with a trivalent impurity, holes are introduced into the valence band. Once again, as for the *n*-type semiconductor previously described, electron-hole pairs will be thermally generated, but holes will always outnumber electrons as the available charge carriers. Holes are now the *majority* carriers, and electrons are the *minority* carriers. A doped semiconductor material having holes as the majority carriers is described as *p*-type.

### 1.10 Extrinsic conduction

In the pure semiconductor material the conductivity was seen to be dependent on its temperature. Conduction due to the thermally generated electron-hole pairs is called *intrinsic* conduction.

In the doped semiconductor material the conductivity is determined largely by the level of doping. Conduction which is possible due to the charge carriers made available by the impurities, is described as *extrinsic* conduction.

### 1.11 Electrical neutrality of *p*-type or *n*-type semiconductor

When a piece of semiconductor is described as *p*-type or *n*-type, this does not indicate that it has any net positive or negative charge. The terms merely indicate which type of charge carrier is predominantly available for conduction when a p.d. is applied.

## Problems

1. (a) Sketch a simple two-dimensional diagram to show the arrangement of the atoms in a sample of pure silicon.
   (b) Explain what is meant by the following:
      (i) covalent bond
      (ii) electron-hole pair
      (iii) intrinsic conduction
2. With the aid of simple diagrams, describe how under the influence of an external p.d., holes in a semiconductor material move towards the negative end of the semiconductor material.
3. Using simple two-dimensional diagrams where appropriate, explain how pure silicon can be treated such that conduction is predominantly due to (i) electrons and (ii) holes.

# Chapter 2
# The *p–n* Junction and its Characteristics

## 2.1 The *p-n* junction

In Chapter 1 it was explained that semiconductor materials can be doped to make them *p*-type or *n*-type. If a piece of semiconductor is doped such that one part of it is *p*-type and the remainder is *n*-type, a *p-n junction* is formed at the interface where the change from *p* to *n* occurs.

Although not a practical manufacturing technique, it is convenient to consider the *p-n* junction being formed by joining together separate pieces of *p*-type and *n*-type material.* The join would, of course, have to be such that there was no break in the lattice structure between the *p* and *n* materials. Figure 2.1 represents the *p-n* junction at the moment when the two parts are 'joined'.

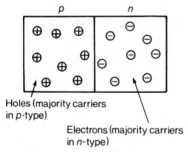

Fig 2.1 Representation of a *p-n* junction at the instant when the two parts are 'joined' (see text). Only majority charge carriers are shown.

## 2.2 Diffusion

Let us consider how the majority charge carriers behave once the junction is formed. The *n*-type side has a high concentration of electrons, whereas the *p*-type side has very few. There is a natural tendency

---

* In practice such processes as the alloying technique or the diffusion technique would be used to produce the *p-n* junction. Such techniques are not dealt with in this book since a knowledge of them is not essential for a basic understanding of the devices they produce.

for the electrons in the *n*-type to spread across into the *p*-type in an attempt to even up the concentrations. This is essentially a non-electrical process similar, for example, to the action of a small drop of ink dispersing itself in water. The phenomenon is called *diffusion*. On entering the *p*-type the electrons encounter a high concentration of holes and recombinations take place.

Similarly the *p*-type side has a high concentration of holes whereas the *n*-type side has very few. Holes from the *p*-type therefore diffuse across into the *n*-type where they recombine with the available electrons. Figure 2.2 illustrates these initial diffusion currents.

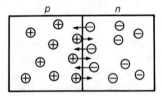

Fig 2.2 Representation of the diffusion of majority charge carriers immediately after the formation of the *p-n* junction.

## 2.3 Depletion layer and barrier potential

The regions close to the junction lose their majority charge carriers because of the diffusion and recombination described above. The thin layer formed by these regions is known as the *depletion* layer (i.e. it is depleted of majority carriers).

In the part of the depletion layer on the *p*-type side of the junction, the electrons acquired by diffusion will negatively ionise* the trivalent impurity atoms by recombining with the holes which these impurity atoms provided (i.e. completing the fourth covalent bonds). Adding this to the fact that this region also loses holes by diffusion, it is clear that this part of the depletion layer must obtain a net negative charge.

In the depletion region on the *n*-type side of the junction, the holes acquired by diffusion will positively ionise the pentavalent impurity atoms by recombining with the electrons which these impurity atoms provided. The result of this, together with the fact that there has been a loss of electrons due to diffusion, is that this part of the depletion region obtains a net positive charge.

A potential difference now exists across the depletion layer because of these acquired charges (these acquired charges are, of course, bound into the semiconductor structure). This p.d. is represented by $V_B$ in Fig. 2.3.

The potential difference builds up to a level where it presents a

---

* An *ion* is an atom which has gained or lost an electron, and as such has had its charge balance upset. An atom with an extra electron is negatively ionised, whilst an atom which has an electron missing is positively ionised.

barrier to further diffusion. In effect, the action is like this. The negative potential established on the *p*-type side will tend to repel electrons that are attempting to diffuse across from the *n*-type. Every electron which succeeds in overcoming this repelling effect, and crosses the

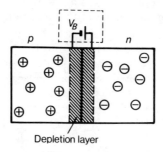

Fig. 2.3 *p-n* junction with $V_B$, representing the barrier potential developed by charge differences in the depletion layer.

junction, increases the negative potential and makes it more difficult for the next electron. Similarly, the positive potential developed on the *n*-type side, builds up to a level which inhibits further diffusion of holes across into the *n*-type. Since the potential difference across the junction is a barrier to the diffusion process, it is quite reasonably called the *barrier potential*. The magnitude of the barrier potential is such that the net charge movement across the junction is zero.*

The representation of the *p-n* junction in Fig. 2.3 gives no indication of just how narrow the depletion layer is in reality. Actual widths vary but are typically less than 1 μm. It should also be noted that the depletion region will not necessarily penetrate by equal amounts on either side of the junction. The actual penetration depends on doping levels which may well be different for the two sides. Higher doping levels will result in less penetration and hence a narrower depletion region.

## 2.4 Forward bias

If an external potential difference is applied to the *p-n* junction such that the *p*-type side is made positive with respect to the *n*-type side (i.e. positive to *p* and negative to *n*) the *p-n* junction is said to be forward biased. Figure 2.4 illustrates this situation.

The forward bias $V_F$, attracts some of the charges which ionised impurity atoms in the depletion region, and so reduces the bound

---

* Notice that the polarity of the barrier potential is such that thermally generated minority carriers (i.e. electrons in *p*-type and holes in *n*-type) will drift across the junction. The barrier potential rises to a level at which this drift current is balanced by the diffusion current so that the net charge movement across the junction is zero when no external p.d. is applied.

charge levels there. That is, the positive terminal of $V_F$ attracts some of the electrons that established a negative charge in the p-type side of the depletion region, and the negative terminal of $V_F$ attracts some of the holes that established a positive charge in the n-type side of the depletion region. This results in both a lowering of the barrier potential and a corresponding narrowing of the depletion layer. Since the bar-

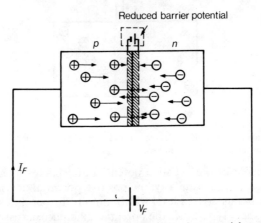

Fig 2.4 Forward biased p-n junction. The reduced barrier potential enables majority charge carriers to diffuse across the junction.

rier potential is reduced, majority carriers can again diffuse across the junction in large numbers. This majority carrier movement across the junction does not rebuild the barrier potential, but results in current in the external circuit. That is, electrons lost by diffusion from the n-type to the p-type, are replaced by electrons from the negative terminal of $V_F$. The electrons gained by the p-type will be given up to the positive terminal of $V_F$. This represents a movement of electrons from the negative to the positive terminal of $V_F$ via the semiconductor device. A similar situation exists for the diffusion of holes from the p-type except that holes as charge carriers are confined to the semiconductor material. The holes diffusing from the p-type are lost by recombination in the n-type, the electrons needed to keep the charge balance being drawn from the negative terminal of $V_F$. So the flow of holes from p to n in the semiconductor again results in electrons being drawn from the negative terminal of $V_F$. The supply of holes in the p-type is maintained by the p-type giving up electrons to the positive terminal of $V_F$. The current in the circuit external to the semiconductor materials is due entirely to electron movements and has a magnitude equal to the sum of the electron and hole currents in the p-n junction.

## 2.5 Reverse bias
If an external potential difference is applied to the p-n junction such that

# The p-n *Junction and its Characteristics*

the *p*-type is made negative with respect to the *n*-type, the *p-n* junction is said to be reverse biased. Figure 2.5 illustrates this situation.

The negative terminal of $V_R$ will attract the holes in the *p*-type and they will be drawn away from the junction. Similarly, the positive terminal of $V_R$ will attract the electrons in the *n*-type drawing them

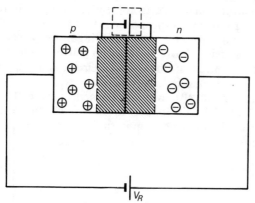

Fig 2.5 Reverse biased *p-n* junction. Since the majority carriers are attracted away from the junction the depletion layer is widened.

away from the junction. Since the majority carriers are attracted away from the junction they are clearly unable to cross it. In effect then, the reverse biased *p-n* junction represents an open circuit to majority carrier currents. It is worth noting, that drawing the majority carriers away from the junction widens the depletion layer as Fig. 2.5 shows. The width of the depletion layer is, therefore, a function of the magnitude of the reverse bias p.d.

Now, with reference to Fig. 2.6, let us consider the behaviour of the minority carriers when the junction is reverse biased.

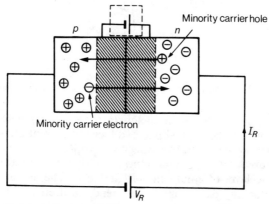

Fig 2.6 Reverse biased *p-n* junction. The thermally generated minority charge carriers are swept across the junction by the barrier potential.

The polarity of the barrier potential is such that the thermally produced minority carriers will be swept across the junction. The minority carrier electrons are gathered by the *n*-type side and the minority carrier holes are gathered by the *p*-type side. These minority carrier movements result in a small current in the circuit. That is, the arrival in the *n*-type of a minority carrier electron from the *p*-type will result in an electron being given up to the positive terminal of $V_R$. Similarly a minority carrier hole from the *n*-type arriving in the *p*-type will necessitate an electron being drawn from the negative terminal of $V_R$. The magnitude of the current in the external circuit is again equal to the sum of these electron and hole currents.

Since there are relatively few minority carriers available, this reverse current is extremely small. Its actual value is of course temperature dependent, since the number of minority carriers produced increases with temperature.

## 2.6 I-V characteristics

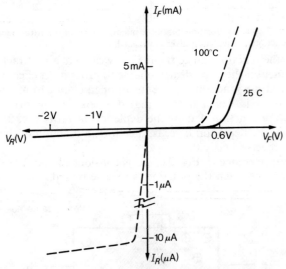

Fig 2.7 *I-V* characteristics of a silicon *p-n* junction at 25°C and 100°C.

Figure 2.7 shows the *I-V* characteristic for a silicon *p-n* junction. When forward biased, the applied p.d. and forward current are exponentially related as the characteristic suggests. Notice that at nominal room temperature (25°C), a p.d. of approximately 0·6 V is required before a reasonable level of conduction is obtained. The actual forward p.d. required for any particular current tends to decrease by about 2 mV for every degree centigrade rise in junction temperature.

As previously described in section 2.5, a small current, normally

## The p-n Junction and its Characteristics

described as the saturation current or leakage current, is to be expected when the junction is reverse biased. This current is due to thermally produced minority carriers. Since, at room temperatures, there are very few minority carriers, a reverse p.d. of only a fraction of a volt is all that is needed to involve all of them in the conduction process. This means that the reverse current reaches a saturation level at a very small reverse p.d. The slight increase in reverse current above this initial saturation level, which is apparent from the curves, is due to charge leakage across the surface of the semiconductor material in the vicinity of the junction. This can be regarded as a high resistance path in shunt with the reverse biased junction. Ignoring surface leakage effects, the actual value of the saturation current is, of course, a function of temperature. For a given p.d., the saturation current in a silicon $p$-$n$ junction approximately doubles for every 10°C rise in junction temperature.

Figure 2.8 shows that for a germanium $p$-$n$ junction, a forward potential of approximately 0·2 V is required to obtain reasonable conduction at room temperature. Comparison of the saturation currents for the germanium and silicon junctions, shows that the reverse current in the germanium device is considerably higher. As with the silicon device however, the actual value of the current approximately doubles for every 10°C rise in junction temperature.

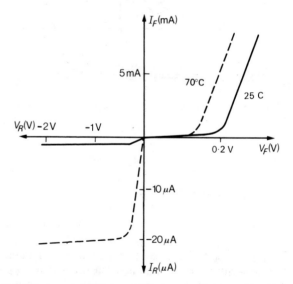

Fig 2.8 $I$-$V$ characteristics of a germanium $p$-$n$ junction at 25°C and 70°C.

### 2.7 Forward resistance

For many electronic devices the d.c. resistance that would be found by dividing the direct voltage across the device, by the resulting direct current, is of little value. The resistance which the device offers to a small *change* in the applied voltage is found to be more useful. This is the resistance that would be encountered by a small a.c. signal when added to a direct voltage applied to the device. This quantity is generally known as the *a.c. resistance*, and its importance will become clear when considering transistor input impedance in Chapter 4.

Consider however, the *p-n* junction characteristic in Fig. 2.9, where a small a.c. signal $v_1$, is added to the forward bias p.d. $V$, applied across the junction.

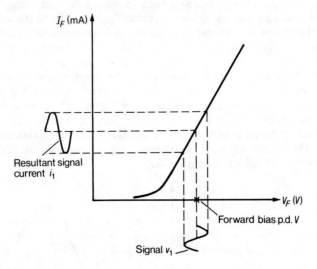

Fig 2.9 *I-V* characteristics of a *p-n* junction. The signal voltage $v_1$ results in the signal component of current $i_1$.

The signal voltage $v_1$ results in the signal current $i_1$. The effective a.c. resistance $r_F$, is therefore:

$$r_F = \frac{v_1}{i_1} \Omega$$

Since the *I-V* characteristic of the junction is an exponential curve, the actual value for $r_F$ obtained in this way, will vary with both the position on the curve about which the signal is applied (i.e. the value of $V$), and the magnitude of the signal component.

## 2.8 Semiconductor diode
The characteristics of the *p-n* junction show it to be a device which passes current of significant magnitude in one direction only (i.e. when it is forward biased). The two terminal device which the *p-n* junction

forms is called a semiconductor junction diode. Its circuit symbol is shown in Fig. 2.10. The *p*-type and *n*-type regions are referred to as the anode and cathode respectively. In order to make the diode conduct, its anode must be positive with respect to its cathode. If the device is a germanium junction diode, a forward p.d. in the region of 0·15 V to

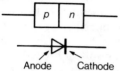

Fig 2.10 *p*-*n* junction and the corresponding semiconductor diode symbol.

0·3 V can be expected when it is conducting. Forward potentials differences of approximately 0·5 V to 0·8 V are common for silicon diodes.

## 2.9 Reverse breakdown

If the reverse p.d. applied to a *p-n* junction is increased to a sufficiently high value, a breakdown effect will occur. This phenomenon is illustrated in Fig. 2.11. At the breakdown p.d. (marked $V_z$ in Fig. 2.11) the reverse current rises sharply to a value limited largely by whatever resistance is in series with the device. In this situation, the power dissipation in the device, calculated as the product of the reverse current and the p.d. $V_z$, may well be in excess of that which the device can safely withstand and it may be irrevocably damaged. In general therefore, it is important that the reverse p.d. across a diode does not reach the breakdown value.

It will however be noticed from Fig. 2.11, that once breakdown has

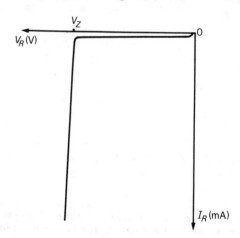

Fig 2.11 Illustrating the reverse breakdown characteristic of a *p-n* junction.

occurred very little further increase in reverse p.d. is required to increase the current to relatively high values. Looking at this in another way, we may say that the p.d. remains almost constant for a large range of currents. Because of this property certain junction diodes, operated in the breakdown region, may be used to provide a voltage reference.* Such diodes, known as *voltage reference diodes* (*VRD*), are specifically designed to be operated in this way (see section 3.8).

The physical process responsible for the breakdown, depends on the actual width of the depletion layer for any given reverse p.d. (i.e. on the doping level). Two mechanisms of breakdown are possible, and these are known as *Zener* breakdown and *avalanche* breakdown.

**Zener breakdown:** This form of breakdown occurs in junctions which, being heavily doped, have narrow depletion layers. At the breakdown p.d. the electric field across this narrow layer is sufficient to break covalent bonds, thereby generating electron-hole pairs. Any further small increase in the reverse p.d. therefore results in a very large increase in the number of charge carriers available. This accounts for the low resistance of the diode in the breakdown region.

**Avalanche breakdown:** It was explained in section 2.5 that a small current exists in a reverse biased *p-n* junction due to minority carriers. As these minority carriers drift through the depletion region they may well be involved in collisions with the semiconductor atoms. If the p.d. across the depletion region is sufficiently high, the minority carriers will acquire sufficient energy to raise other electrons to the conduction band as a result of the energy transferred during these collisions. This means that more electron-hole pairs are generated. In a fairly wide depletion region, a minority carrier may be involved in several such collisions. The newly generated charge carriers themselves drift under the influence of the reverse p.d. and they are likewise involved in collisions producing even more electron-hole pairs. This leads to an avalanche of charge carriers and consequently a very low reverse resistance. Avalanche breakdown therefore occurs in junctions which, being lightly doped, have wide depletion regions.

The actual cause of breakdown in a particular *VRD* may be partly due to the Zener effect and partly due to avalanche. In general the Zener effect tends to predominate in devices which breakdown below about six or seven volts. The avalanche effect predominates in devices which break down at higher potentials. The breakdown p.d. for the Zener effect tends to decrease slightly with increased temperature, whereas the p.d. for the avalanche effect tends to increase slightly.

---

* A voltage reference is an accurately known p.d. from which other circuits can be supplied or to which other potentials can be compared. Its value should ideally remain constant despite variations in temperature, supply potentials or loading.

# The p-n Junction and its Characteristics

## Problems
(Answers at end of book)

1. With reference to a *p-n* junction explain:
   (*i*) how the initial diffusion of majority carriers results in the formation of a barrier potential;
   (*ii*) why as a result of the barrier potential the net diffusion of charge carriers across the junction falls to zero;
   (*iii*) why the region close to the junction is depleted of majority carriers.

2. (*a*) Explain why the application of a forward bias p.d. to a *p-n* junction, results in majority charge movement across the junction, and current in the external circuit.
   (*b*) Explain why the current in a reverse biased *p-n* junction is extremely small. Why does this current reach a saturation level at a very small reverse p.d.?

3. (*a*) Sketch the forward and reverse *I-V* characteristics for (*i*) a silicon diode and (*ii*) a germanium diode. State how the forward bias voltages and leakage currents of the two devices are affected by changes in temperature.
   (*b*) A particular silicon diode has a leakage current of 20 nA at 25°C. Estimate its leakage current at 55°C.

4. Sketch the breakdown characteristic of a voltage reference diode. Briefly describe each of the two physical mechanisms that could be responsible for the breakdown. State which of the two mechanisms is more likely for a diode which breaks down at (*i*) 5 V and (*ii*) 30 V.

# Chapter 3
# Rectification and Simple Power Supply Circuits

## 3.1 Rectification
The electricity supply to industrial and domestic consumers is normally in an alternating voltage form. However, the majority of electronic equipments require direct voltage supplies. This makes it necessary to obtain a direct voltage from an alternating voltage. An essential step involved in achieving this is *rectification*, which is the process of obtaining a *unidirectional* voltage (i.e. one which does not alternate), from an alternating voltage.

## 3.2 Half-wave rectification
Figure 3.1 illustrates the process of *half-wave* rectification. It can be

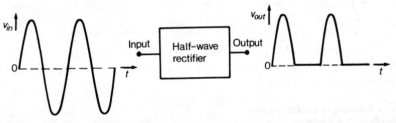

Fig 3.1 Illustrating the process of half-wave rectification. The negative half-cycles have been eliminated in the output waveform.

seen that the negative half-cycles of the input have been eliminated and therefore the output voltage is unidirectional. Half-wave rectification is also achieved if the rectifier circuit eliminates the positive half-cycles thus providing an output of negative half-cycles only. This is illustrated in Fig. 3.2.

Rectification is usually performed by semiconductor diodes (thermionic diodes can be used but semiconductor devices are generally preferred, see section 8.1). It was explained in Chapter 2 that a semiconductor diode will only pass current (of any significant magnitude) when it is forward biased. Figure 3.3 shows a semiconductor diode being used as a half-wave rectifier. On positive half-cycles of the input

# Rectification and Simple Power Supply Circuits

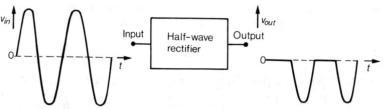

Fig 3.2 Illustrating the process of half-wave rectification. The positive half-cycles have been eliminated in the output waveform.

the anode of the diode is made positive with respect to its cathode and it conducts. This effectively connects the load to the input terminals and hence the positive half-cycles are developed at the output. On negative half-cycles, the anode of the diode is made negative with respect to its cathode. The diode is now reverse biased and will not conduct. The p.d. at the output terminals during the negative half-cycles is therefore zero.

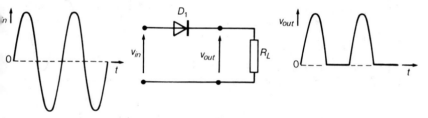

Fig. 3.3 Semiconductor diode being used as a half-wave rectifier.

If the anode and the cathode connections are interchanged it will result in an output of negative half-cycles only (i.e. an output of the form illustrated in Fig. 3.2).

## 3.3 Full-wave rectification

Figure 3.4 illustrates the process of *full-wave* rectification. Notice that

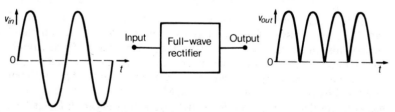

Fig 3.4 Illustrating the process of full-wave rectification. Both the positive and negative half-cycles of the input produce positive half-cycles at the output.

both the positive and negative half-cycles of the input produce positive going half-cycles at the output. Alternatively a full-wave rectifier circuit can be arranged such that a continuous succession of negative going half-cycles appear at the output as shown in Fig. 3.5.

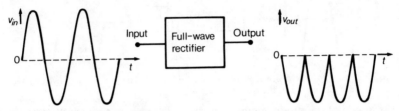

Fig 3.5 Illustrates the process of full-wave rectification with negative half-cycles at the output.

A commonly used full-wave rectifier circuit is one which uses four diodes to form a *bridge*. This circuit arrangement known as a *bridge rectifier* is shown in Fig. 3.6.

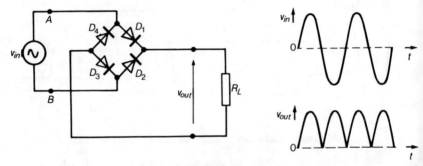

Fig 3.6 Bridge rectifier circuit.

On half-cycles of the input which make point $A$ positive with respect to point $B$, diodes $D_1$ and $D_3$ are forward biased, whereas $D_2$ and $D_4$ are reverse biased. The load current is therefore conveyed via $D_1$ and $D_3$ as shown in Fig. 3.7. $D_4$ is obviously reverse biased since its cathode is connected to point $A$, which is the most positive point in the circuit for this half-cycle (clearly then, at any time during this half-cycle, it is not possible for its anode to be more positive than point $A$ and so this diode cannot be forward biased). Similarly $D_2$ is reverse biased since its anode is connected to point $B$ which is the most negative point during this half-cycle (in this instance it is not possible for its cathode to be more negative, which means it cannot be forward biased). This leaves $D_1$ and $D_3$ which, now that $D_2$ and $D_4$ can be eliminated, can be seen to be in series with the load $R_L$. By following the path from $A$ to $B$ it is clear that the anodes of both of these diodes are directed towards the more positive terminal and hence they are forward biased.

Now consider the alternate half-cycles, that is, those which make

# Rectification and Simple Power Supply Circuits

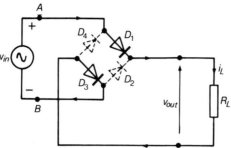

Fig 3.7 Illustrating the current path when *A* is positive with respect to *B*. Diodes shown dotted are reverse biased.

point *B* positive with respect to point *A*. During these half-cycles, $D_2$ and $D_4$ are forward biased and $D_1$ and $D_3$ are reverse biased. The current is therefore developed in the load via $D_2$ and $D_4$. This situation is illustrated in Fig. 3.8.

Inspection of Fig. 3.7 and Fig. 3.8, shows that the load current in $R_L$ has the same direction regardless of the polarity of the input half-cycle. Hence, both the positive and the negative half-cycles of the input produce positive half-cycles at the output as indicated earlier.

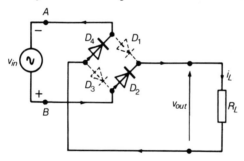

Fig 3.8 Illustrating the current path when *B* is positive with respect to *A*.

An alternative full-wave rectifier arrangement which includes a centre-tapped transformer is shown in Fig. 3.9. For convenience as-

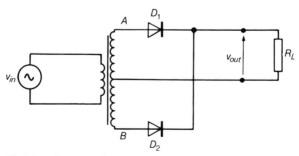

Fig 3.9 Full wave rectifier circuit utilising centre-tapped transformer.

sume that the total number of turns on the secondary winding of the transformer, is twice the number on the primary winding. The total voltage induced in the secondary winding will then be twice the input voltage. Now consider half-cycles of the input which make end *A* of the secondary winding positive with respect to *B*. The centre-tap of the secondary winding will clearly be negative with respect to point *A* by an amount equal in magnitude to the input voltage. The centre-tap will however be positive with respect to point *B* (point *B* is the most negative point on the secondary winding for this half-cycle), once again by an amount equal in magnitude to the input voltage. In other words the voltages at the two ends of the secondary winding are of equal magnitude but antiphase. The two halves of the induced secondary voltage are represented in Fig. 3.10 as $v_{2a}$ and $v_{2b}$.

It can be seen from the diagram that the anode of $D_2$ is connected to

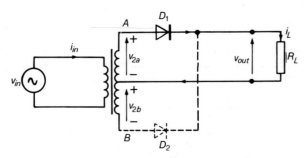

Fig 3.10 Illustrating the current path for the input half-cycle which makes end *A* of the secondary winding positive with respect to *B*. $D_2$ is shown dotted because it is reverse biased.

the most negative point on the secondary winding and, accordingly, it is reverse biased. $D_1$, however, has its anode connected to the most positive point in the circuit and is forward biased. A complete circuit exists through the upper half of the transformer secondary winding via

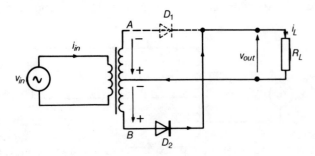

Fig 3.11 Illustrating the current path for the input half-cycle which makes end *B* of the secondary winding positive with respect to *A*.

$D_1$ and the load $R_L$. In effect therefore, a voltage of magnitude equal to the input voltage, is applied to the load via $D_1$.

On alternate half-cycles, end $B$ of the transformer secondary winding is positive with respect to $A$. The polarities of the two halves of the secondary voltage which now apply are shown in Fig. 3.11. It is clear from this diagram that $D_1$ is now reverse biased and $D_2$ is forward biased. A complete circuit exists through the lower half of the secondary winding via $D_2$ and the load $R_L$. Once again, a voltage of magnitude equal to the input voltage is applied via a diode to the load. Examination of Fig. 3.10 and Fig. 3.11, shows that the load current in $R_L$ has the same direction regardless of the polarity of the input half-cycle. Like the bridge rectifier therefore, the output is a continuous series of positive half-cycles.

## 3.4 Reservoir capacitor

The process of rectification provides a unidirectional output voltage but not a steady direct voltage. In particular, the rectified waveforms regularly fall to zero. The addition of a large value *reservoir capacitor* across the output terminals of the rectifier circuit will prevent the output falling to low levels. Consider the circuit in Fig. 3.12, which shows a reservoir capacitor connected across the output of a half-wave rec-

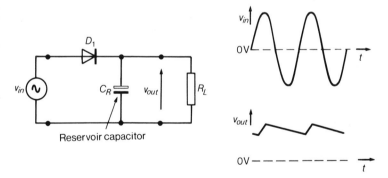

Fig 3.12 Half-wave rectifier circuit with reservoir capacitor.

tifier circuit. When the diode conducts, charge is added to the capacitor. When the diode is reverse biased, the capacitor discharges via $R_L$ and thereby maintains the load current. This action prevents the output voltage from falling to zero. That is, the cathode of the diode is held positive by the capacitor. Now, since the diode can only conduct when its anode is more positive than its cathode, this means that it conducts for only a short time during the positive half-cycles. The portion of the input cycle for which the diode conducts is indicated in Fig. 3.13. The output from the circuit can be regarded as a direct voltage with a *sawtooth* alternating component (usually called the

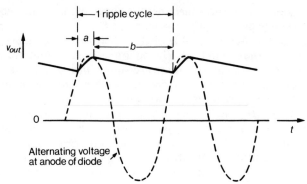

*a:* Input voltage exceeds output voltage and diode conducts charging the capacitor.

*b:* Input voltage falls below output voltage and the capacitor discharges into the load.

Fig 3.13 Output waveform of half-wave rectifier circuit with a reservoir capacitor and resistive load.

*ripple*), superimposed upon it. These components are identified in Fig. 3.14. When the reservoir capacitor is discharged more rapidly by an increased load current, the direct output voltage will decrease and the

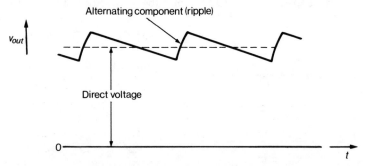

Fig 3.14 Waveform diagram of the output from a rectifier circuit with a reservoir capacitor. This can be considered as a direct voltage with an alternating component superimposed upon it.

ripple amplitude will increase. In order to keep the ripple component as small as possible and to maintain a high direct voltage, the charge lost by the capacitor when the diode is not conducting, must be only a small proportion of the total charge stored. This means that the reservoir capacitor must have a large capacitance.* In order to achieve a large capacitance in a reasonably small physical volume, *electrolytic* capacitors are generally used.

Inspection of the waveform diagram of Fig. 3.13 shows that when a

---

* For power supplies providing outputs of 10 V to 50 V, reservoir capacitors in the range say, 200 $\mu$F to 2000 $\mu$F may be used depending on the current to be delivered.

## Rectification and Simple Power Supply Circuits

reservoir capacitor is added to a half-wave rectifier circuit, the frequency of the ripple is equal to the input supply frequency. Figure 3.15 shows the ripple waveform which results when a reservoir capacitor is added to a full-wave rectifier. It is clear from the diagram that in this case the ripple frequency is twice the supply frequency (e.g. with a supply frequency of 50 Hz, the ripple frequency is 100 Hz). Since the full-wave rectifier replenishes the charge on the reservoir capacitor twice as often as the half-wave rectifier, a higher direct component, with a smaller ripple amplitude is obtained.

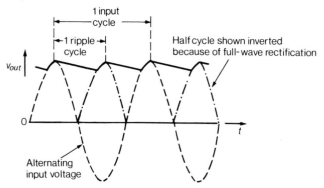

Fig. 3.15 Waveform diagram showing the ripple obtained when a reservoir capacitor is added to a full-wave rectifier circuit.

### 3.5 Maximum reverse voltage

The greatest reverse voltage that a diode can safely withstand is commonly quoted as its *peak inverse voltage* rating (p.i.v. rating). If the load $R_L$ is removed from the circuit in Fig. 3.12, the reservoir capacitor cannot discharge and the output will remain at a level approximately equal to the peak value of the input voltage. Under these circumstances the diode must withstand a maximum reverse voltage equal to twice the peak value of the input. This is because the cathode of the diode will be held at the positive output voltage whilst, on negative half-cycles of the input, its anode will be taken to the negative peak value. It is essential therefore, when choosing a diode for this application, to ensure that its p.i.v. rating is in excess of twice the peak value of the input voltage. That is, with a peak input voltage of say 100 V, a diode with a p.i.v. rating in excess of 200 V is required.

### 3.6 Peak rectifier currents

It was explained in section 3.4 that when a reservoir capacitor is added to a half-wave rectifier circuit, the diode conducts for a short period during one half-cycle. In this short conduction period, the input current must replenish the charge lost by the capacitor during the rest of the cycle. The delivery of a large amount of charge in a short period means a high current. A diode chosen for this application must be capable of carrying this current on a repetitive basis. Furthermore, at the instant when the circuit is first switched on, the reservoir capacitor is completely uncharged. A large *surge* current will therefore flow for what, relatively speaking, is a long period as the capacitor acquires its initial charge. The increase of junction temperature caused by this current, may be sufficient to damage the diode. Because of these considerations, it may be necessary to include a small resistor of say 1 Ω to 5 Ω in series with the diode to limit the maximum current to a safe level. These problems also exist for the diodes in a full-wave rectifier circuit. However, when considering the repetitive current pulses, the average charge to be delivered by any diode will be less since the capacitor is replenished twice as frequently.

## 3.7 Filter circuit

The ripple component of the output p.d. developed across the load can be reduced by connecting a low pass filter (commonly called a smoothing filter) between the reservoir capacitor and the load. A typical low pass filter circuit is shown in Fig. 3.16. Assume $R_F$ is a 220 Ω resistor,

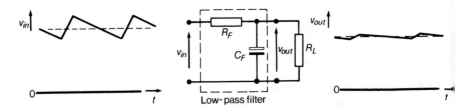

Fig. 3.16 Low-pass filter circuit.

and $C_F$ is a 1000 μF capacitor. The effect of the filter circuit on the ripple component can be illustrated by determining the proportion of the ripple that is developed across $C_F$. Consider a ripple voltage with a frequency of 100 Hz (this is the ripple frequency that would be obtained with a 50 Hz input and a full-wave rectifier circuit). This ripple voltage can be considered to contain many harmonically related sinusoidal components. The most significant of these is a fundamental sine wave with a frequency equal to that of the sawtooth, i.e. 100 Hz. In order to simplify the calculations we shall adopt an approximate approach and ignore all frequency components other than the fundamental. That is, we shall make the assumption that the ripple is a sinusoidal voltage of frequency 100 Hz. The reactance of the capacitor $C_F$, to the ripple is therefore:

$$X_c = \frac{1}{2\pi f C_F}$$

# Rectification and Simple Power Supply Circuits

$$= \frac{1}{2\pi \times 100 \times 1000 \times 10^{-6}} \, \Omega$$

$$= \frac{10}{2\pi} \, \Omega$$

$$= 1.592 \, \Omega$$

The magnitude of the ripple component across $C_1$ can be found using a potential divider approach. That is:

$$V_{o/p\,(\text{RIPPLE})} = \frac{X_c}{\sqrt{R_F^2 + X_c^2}} \times V_{i/p\,(\text{RIPPLE})}$$

where $V_{o/p\,(\text{RIPPLE})}$ is the amplitude of the ripple developed across $C_F$. Note that this expression ignores the shunting effect of $R_L$ on $C_F$ since this will be negligible.

If the amplitude of the input ripple is 1 V, then:

$$V_{o/p\,(\text{RIPPLE})} = \frac{1.592}{\sqrt{220^2 + 1.592^2}} \times 1 \text{ V}$$

$$\simeq \frac{1.592}{220} \text{ V}$$

$$\simeq 0.00724 \text{ V}$$

$$\simeq 7.24 \text{ mV}$$

The filter circuit has therefore reduced the ripple from an amplitude of 1 V to an amplitude of approximately 7 mV.

The filter will also reduce the direct voltage component, but this reduction will be less significant since $C_F$ will be open circuit to d.c. The direct component of the output voltage will be determined by potential divider action between $R_F$ and $R_L$, where:

$$V_{o/p\,(\text{d.c.})} = \frac{R_L}{R_F + R_L} \times V_{I/P\,(\text{d.c.})}$$

If for our example $R_L$ is 1 kΩ, and $V_{I/P\,(\text{d.c.})}$ is 20 V, then:

$$V_{o/p\,(\text{d.c.})} = \frac{1000}{220 + 1000} \times 20 \text{ V}$$

$$= 16.4 \text{ V}$$

The results in this example can be summarised as follows. The d.c. component of the output has been reduced by approximately 18% by the inclusion of the filter circuit, whereas the ripple component has been reduced by approximately 99.3%.

The reduction in the direct component can be almost eliminated if a large value inductor is used in place of $R_F$. This will have a large reactance to the ripple but very little resistance and so will not significantly affect the direct voltage. Large value inductors are however very expensive, bulky and heavy, and so are not often used. This ripple filtering action can also be achieved electronically by means, for example, of a voltage regulator such as that described in section 3.8.

A complete low voltage power supply, incorporating a step-down transformer to reduce the a.c. input voltage, is shown in Fig. 3.17. $C_R$ is the reservoir capacitor, and $R_F$, $C_F$ form a low pass filter circuit. $R_1$ is included to limit the maximum current in the diodes to a safe level. The fuse $F_1$ will blow if a fault occurs which results in excessive current being drawn.

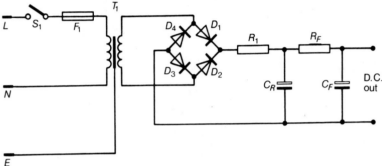

Fig 3.17 A mains derived d.c. power supply circuit.

## 3.8 Simple voltage stabilizer circuit

The output voltage from a circuit of the type shown in Fig. 3.17 will decrease as the load current delivered is increased. If the output voltage is required to remain substantially constant for a range of load currents, a voltage stabiliser circuit can be used. This circuit makes use of a voltage reference diode (*VRD*). The principle of the *VRD* is described in section 2.9 but for ease of reference its *I–V* characteristic is redrawn in Fig. 3.18. Note that once the *VRD* is operating in the breakdown region, the p.d. across it remains almost constant regardless of the current it carries. The very small increase in this p.d. which does occur as the current is increased can be determined by considering the slope resistance of the device. Consider the portion of the characteristic in the vicinity of the point *P*, this has been enlarged and redrawn in Fig. 3.19.

The slope resistance, $$r_2 = \frac{\Delta V}{\Delta I} \, \Omega$$

Hence, if $r_2$ is known, the increase in voltage $\Delta V$ for a given increase in current $\Delta I$, can be found using the above expression:

i.e. $$\Delta V = \Delta I r_z$$

# Rectification and Simple Power Supply Circuits

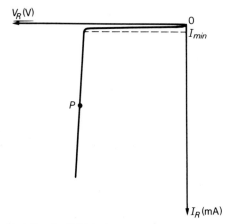

Fig 3.18 *I-V* characteristic of a *VRD* (only reverse characteristic shown).

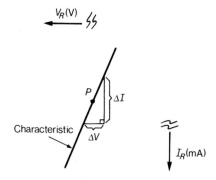

Fig 3.19 Enlarged portion of the *I-V* characteristic of the *VRD*.

Slope resistances as low as 5 Ω or less are typical, and for our purposes we shall consider $r_z$ to be negligible. That is, we shall assume there is no perceptible change in p.d. across the *VRD* for the range of currents considered.

Consider then the simple voltage stabiliser circuit shown in Fig. 3.20. The direct input voltage $V_{IN}$, must be larger than the breakdown

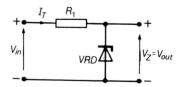

Fig 3.20 Simple voltage stabiliser circuit. Note that the 'cathode' end of the diode is positive with respect to the 'anode' end since the device is operating in the reverse breakdown region.

p.d. of the *VRD*, $V_z$. Note that the output voltage is taken from across the *VRD* and hence the output p.d. is equal to $V_z$. The difference between $V_{IN}$ and $V_z$ is developed across $R_1$. $R_1$ is chosen to limit the current $I_T$, to a level which will not result in excessive power dissipation in the *VRD* when the circuit is not loaded.

In Fig. 3.21 a load resistor $R_L$ has been connected across the output terminals. Let us assume that the addition of $R_L$ does not change $V_z$ (whether or not this assumption is reasonable will be seen later). If $V_z$ has not changed, then the p.d. across $R_1$ remains unaltered. The current in $R_1$ is not therefore affected by the addition of $R_L$, and remains unchanged. Since the load is carrying a current $I_L$, and the input current $I_T$ has not changed, this must mean that the *VRD* current has reduced by an amount equal to $I_L$. Looking at the characteristic of the *VRD* given in Fig. 3.18, we can see that providing a certain minimum current $I_{MIN}$ exists in the device, the p.d. across its terminals will remain at approximately $V_z$. If then, when the current in the *VRD* is reduced by an amount equal to $I_L$, this still leaves a current of at least $I_{MIN}$ in the *VRD*, the p.d. across the device will remain at $V_z$ (remember we are assuming $r_z$ to be negligible). It is quite reasonable therefore, to assume that the output voltage will remain at $V_z$ for a range of load currents up to a maximum of $I_T - I_{MIN}$.

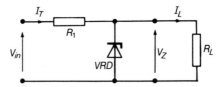

Fig 3.21 Simple voltage stabiliser circuit with load resistor $R_L$.

Let us examine the principle of this circuit again, this time introducing some practical values. Consider a 10 V *VRD* having negligible slope resistance and a maximum power dissipation of 500 mW. This device is to be used to provide a stabilised 10 V output from an unregulated d.c. supply having a maximum voltage of 15 V. The *I–V* characteristics of the device, and the circuit in which it is to be used, are shown in Fig. 3.22(*a*) and Fig. 3.22(*b*) respectively. $R_1$ will be chosen to limit the power dissipation in the *VRD* to 500 mW when the maximum input voltage of 15 V is present (and the circuit is unloaded as shown). This will permit the largest possible load current to be delivered by the circuit. Now the power dissipated in the *VRD* can be found using the expression: $P_z = I_z V_z$.

Therefore, the maximum current that the *VRD* can safely carry is:

$$I_{z(MAX)} = \frac{P_{z(MAX)}}{V_z} \qquad \text{(where } P_{z(MAX)} \text{ is the maximum } VRD \text{ power dissipation)}$$

# Rectification and Simple Power Supply Circuits

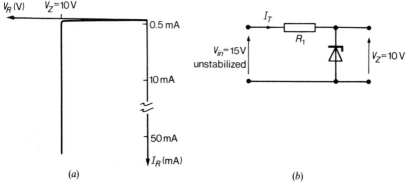

Fig 3.22: (a) I-V characteristics of a VRD.
(b) Simple voltage stabiliser circuit incorporating the VRD.

$$\therefore \quad I_{z(MAX)} = \frac{500 \times 10^{-3}}{10} \text{ A}$$

$$= 50 \text{ mA}$$

Now the p.d. across $R_1$ is the difference between $V_{IN}$ and $V_z$, i.e.:

$$V_{R_1} = V_{IN} - V_z$$

$\therefore$ when $V_{IN}$ is 15 V:

$$V_{R_1} = 15 - 10 \text{ V}$$
$$= 5 \text{ V}$$

now $\quad R_1 = \dfrac{V_{R_1}}{I_T}$

and when the circuit is unloaded $I_T = I_{z(MAX)}$

$$\therefore \quad R_1 = \frac{V_{R_1}}{I_{z(MAX)}}$$

$$= \frac{5}{50 \times 10^{-3}} \Omega$$

$$= 100 \ \Omega$$

The unloaded circuit showing the calculated value of $R_1$ and the magnitude of $I_T$, is shown in Fig. 3.23(a). If now, as shown in Fig. 3.23(b), a 250 Ω load resistor is connected across the output terminals of the circuit, it will take a load current:

$$I_L = \frac{V_z}{R_L}$$

Now we shall assume, once again, that $V_z$ remains at 10 V. Whether or not this is the case will be seen from the *I–V* characteristic of the *VRD*, once the current remaining in the *VRD* has been determined.

$$\therefore \quad I_L = \frac{10}{250} \text{ A}$$
$$= 40 \text{ mA}$$

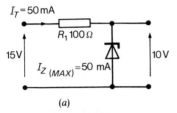

Fig 3.23 (*a*) Voltage stabiliser circuit unloaded.

(*b*) Voltage stabiliser circuit with 250 Ω load.

Now if the output p.d. remains at 10 V, the p.d. across $R_1$ remains at 5 V, and the current $I_T$ is still 50 mA. Clearly then, if the load current is 40 mA the current in the *VRD* is 10 mA. Examination of the characteristic of the *VRD* given in Fig. 3.22(*a*), shows that when the device carries a current of 10 mA, the p.d. across its terminals is still 10 V. The assumption made earlier is therefore justified. In fact, from the characteristic of Fig. 3.22(*a*) it would appear that the device will maintain a terminal p.d. of 10 V providing it carries a minimum current of 0·5 mA. This suggests that the circuit in Fig. 3.22(*b*) can deliver a maximum current of 49·5 mA at 10 V.

Circuits of this type will also maintain a constant output voltage even though there may be variations in the input voltage. If the circuit is supplied from a rectifier and reservoir capacitor arrangement, there will be variations in the input voltage due to the ripple. In order to examine the effect of changes in the input, consider the situation in Fig. 3.24 where the input voltage to the circuit has fallen to 13 V.

Now, the p.d. across $R_1$:

$$V_{R_1} = V_{IN} - V_z$$

and so, assuming $V_z$ still to be equal to 10 V:

$$V_{R_1} = 13 - 10$$
$$= 3 \text{ V}$$

The current in $R_1$:
$$I_T = \frac{V_{R_1}}{R_1}$$
$$= \frac{3}{100} \text{ A}$$
$$= 30 \text{ mA}$$

## Rectification and Simple Power Supply Circuits

The load resistor $R_L$ will take a current:

$$I_L = \frac{V_z}{R_L}$$

where for the circuit of Fig. 3.24, $R_L$ is 400 Ω

∴ $$I_L = \frac{10}{400} \text{ A}$$

$$= 25 \text{ mA}$$

The load takes 25 mA of the total input current of 30 mA. The current in the $VRD$ is therefore 5 mA, which, as the characteristic of Fig. 3.22(a) indicates is more than sufficient to keep the device in the breakdown region. Hence even though the input voltage to the circuit has fallen, the output voltage remains constant at 10 V. Note however, that the output voltage will decrease if the input voltage falls below the level necessary to maintain the minimum breakdown current in the $VRD$.

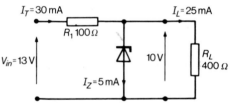

Fig 3.24 Circuit of Fig 3.23(a) with reduced input voltage and 400 Ω load.

A complete circuit which uses a step-down transformer to reduce the mains supply to the required level, and a $VRD$ circuit to stabilise the output voltage is shown in Fig. 3.25. This circuit still includes a reservoir capacitor $C_R$, since the $VRD$ stabiliser circuit cannot function

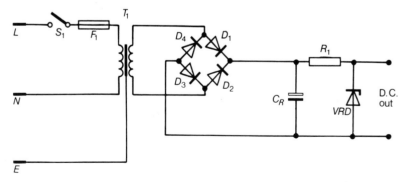

Fig 3.25 A mains derived d.c. power supply which includes a $VRD$ voltage stabilising circuit.

unless the input voltage is held up above a certain minimum level. The filter circuit discussed in section 3.7 is not however required. This is because the ripple voltage at the reservoir capacitor represents a variation in the input voltage to the stabiliser circuit. It has already been shown that such changes will not cause any significant change in the p.d. at the output. Hence in addition to stabilising the d.c. output voltage, the *VRD* circuit also operates as a very effective ripple filter. Finally, as for Fig. 3.17, it may be necessary to include an additional resistor to limit the maximum diode currents (see $R_1$ in Fig. 3.17).

## Problems
(Answers at end of book)

1. (a) Draw the circuit of a bridge rectifier with a resistive load. Indicate on the diagram the current paths for positive and negative half-cycles of the input.
   (b) State the maximum reverse voltage across each diode in terms of the peak value of the a.c. input voltage.

2. (a) Draw the circuit of a full-wave rectifier that uses a centre-tapped transformer and two diodes. If the circuit has a resistive load, indicate the current path for the positive and negative half-cycles of the input.
   (b) If the transformer has a 1:2 (step up) turns ratio, state the maximum reverse voltage across each diode in terms of the peak value of the a.c. input voltage.

3. (a) Explain how the addition of a reservoir capacitor across the output of a rectifier circuit can prevent the output voltage falling to low levels. Sketch the output voltage waveform and identify the direct voltage and ripple component.
   (b) Explain why a small value resistor is often connected in series with a rectifier circuit which is feeding a reservoir capacitor.

4. A voltage stabiliser circuit of the type shown in Fig. 3.20 is supplied from a direct voltage of 14 V. The *VRD* has a breakdown voltage of 8 V, negligible slope resistance, and a maximum power dissipation of 600 mW. Determine:
   (i) the minimum resistance for $R_1$;
   (ii) the approximate maximum load current that this circuit can deliver at 8 V.

5. (a) A series circuit comprising a 5 V *VRD*, an 18 V *VRD* and a 120 Ω resistor, is connected across a 30 V supply to produce a stabilised output of 23 V. Assuming the *VRD*'s have negligible slope resistances, determine:
   (i) the power dissipated in the 18V *VRD* under no load conditions;
   (ii) the approximate maximum output current that can be delivered at 23 V.
   (b) State a possible advantage of using the two *VRD*'s in series rather than a single device.

# Chapter 4
# The Bipolar Transistor

**4.1 Basic construction**

The properties of the *p-n* junction were described in Chapter 2. The bipolar transistor has three distinct semiconductor regions which form two *p-n* junctions. Three connections are made to the device, one to each of these semiconductor regions. The physical construction is such that one of the three connections can be used to control the current in the other two.

Figure 4.1 shows the two possible arrangements for the bipolar transistor, and their corresponding circuit symbols.

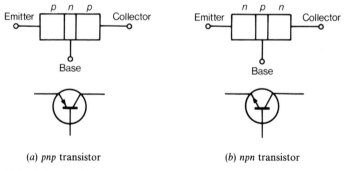

Fig 4.1 Basic forms of construction for bipolar transistors and their corresponding circuit symbols.

The two forms of construction are clearly identified by the *pnp* and *npn* descriptions. In both cases the three semiconductor regions are called the *emitter*, *base* and *collector* as shown. Note that one *p-n* junction is formed between the base and emitter, and the second between the collector and base.

For reasons which will become clear, a practical device has a base region which is extremely thin (typically $10^{-5}$ m), and very lightly doped compared with the other two regions. Furthermore, the emitter is more heavily doped than the collector.

## 4.2 Principle of operation

We will first examine the principle of operation of the *npn* transistor. Consider Fig. 4.2 in which the emitter of the *npn* transistor has been left open circuit. For simplicity the majority carriers have been omitted from this diagram; only the thermally generated minority carriers in the collector and base are shown.

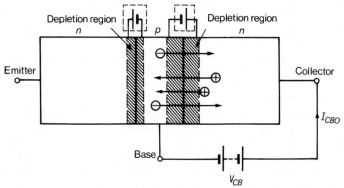

Fig 4.2 *npn* transistor with its emitter open circuit. Only thermally generated minority charge carriers are shown.

The collector-base junction is reverse biased by $V_{CB}$ and so only the minority carriers will be able to cross it. The current in the collector, which under these circumstances is designated $I_{CBO}$ (the collector base leakage current with emitter open circuit), will therefore be very small. The actual magnitude of $I_{CBO}$ will, of course, be a function of temperature, since the number of minority carriers available is temperature dependent. Remember that minority carriers are swept across a reverse biased junction by the barrier potential which exists in the depletion region (see section 2.5).

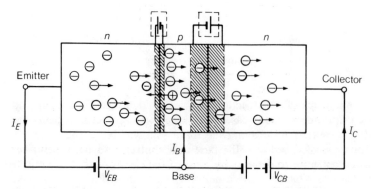

Fig 4.3 Transistor action. The majority of the electrons injected into the base from the *n*-type emitter, are swept across into the collector by the collector-base barrier potential.

# The Bipolar Transistor

Now consider Fig. 4.3 in which the collector-base junction is still reverse biased, but the emitter-base junction is forward biased by $V_{EB}$. This forward bias will enable majority carriers to cross the emitter-base junction. That is, holes will cross from the base to the emitter, and electrons will cross from the emitter to the base. Once electrons cross from the emitter into the base, they become indistinguishable from the minority carrier electrons represented in Fig. 4.2. This means that it is very likely that they will be swept across into the collector region by the collector-base barrier potential. The physical design of the transistor is arranged such that *most* of the electrons which are injected from the emitter into the base are picked up by the collector in this way. A collector current therefore results which, disregarding $I_{CBO}$, is directly dependent on the magnitude of the emitter current.

The emitter current has two components. One component due to electrons crossing from emitter to base, and the second due to holes crossing from base to emitter. Because the base is lightly doped, the component due to holes is a very small proportion of the total.* The emitter current is therefore mainly due to electrons from the emitter region. Since, as stated above, the collector picks up the majority of the electrons which are injected from the emitter into the base, it is clear that the collector current must be almost equal to the emitter current. In a typical device, the collector current may be say, 99% of the emitter current. The 1% difference between these two currents is of course, the base current. In a practical situation therefore, if the emitter current is set to 1 mA, the collector current might be 0·99 mA and the base current 0·01 mA. If the emitter current is increased to 2 mA, the corresponding collector and base currents would be 1·98 mA and 0·02 mA respectively (i.e. increased by the same proportion). Figure 4.4 illustrates the first of these two examples.

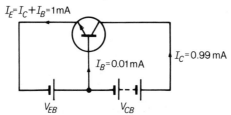

Fig 4.4 Circuit indicating the division of emitter current between collector and base. (Note: Arrow on emitter of circuit symbol points in direction of conventional current in the emitter circuit.)

Having established the basic action of the *npn* transistor, it is worth considering in more detail, why most of the electrons that are injected

---

* It is desirable to keep the component of the emitter current which is due to holes, as small as possible. This is because this component does not contribute to transistor action since no collector current is derived directly from it.

into the base are picked up by the collector. The concentration of electrons in the base is clearly greatest near the emitter because they originate from there. Consequently they tend to diffuse across the base region towards the collector despite the fact that the p.d. between base and emitter is encouraging them to drift towards the base terminal. Since the base is very thin, most of them will diffuse to the collector side of the base region before they can reach the base terminal. When they enter the collector depletion region, the collector-base barrier potential will sweep them across into the collector. Only a few electrons are lost by recombination with holes as they cross the base because it is very lightly doped. Those that are lost in this way result in a component of base current which merely compensates the loss of holes. So as suggested earlier, most of the electrons from the emitter are picked up by the collector.

It is important to recognise that the actual proportion of the electrons from the emitter which enter the collector (and therefore the collector current), is largely independent of the magnitude of the collector-base p.d. $V_{CB}$. This is because the collector-base carrier potential will, almost regardless of its actual value, act upon those electrons which arrive at the collector side of the base region, sweeping them across the collector-base junction. Hence, it is the proportion of the injected electrons which reach the collector side of the base, rather than $V_{CB}$, which determines the collector current.

In order to produce transistor action in a *pnp* transistor, the polarities of both $V_{CB}$ and $V_{EB}$ will have to be the reverse of those shown in Fig. 4.3. The principle of operation is similar to that described for the *npn* transistor except that in this case, the emitter will inject holes into the base which will be swept across the collector-base depletion region, producing the collector current. The supply polarities and the direction of the conventional currents in the electrodes of a *pnp* transistor are shown in Fig. 4.5.

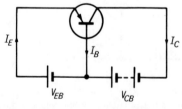

Fig 4.5 Supply polarities and directions of conventional currents for a *pnp* transistor.

## 4.3 Basic principle of a common base amplifier

It was explained in section 4.2 that the collector current is largely independent of the collector-base p.d., but is directly dependent on the emitter current. Consider the circuit of Fig. 4.6 which incorporates an *npn* transistor, the function of which is to amplify the input signal, $v_i$.

# The Bipolar Transistor

The emitter-base junction is forward biased by $V_{EB}'$ and the collector supply is provided by $V_{CC}$. These potentials will establish a steady collector current.

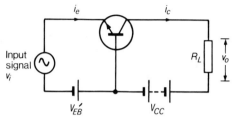

Fig 4.6 Simple common base amplifier circuit. Note: $i_e$ and $i_c$ are signal components of the emitter and collector currents respectively.

This is not a practical circuit, but it illustrates the conditions required for correct operation. The input signal is connected between the emitter and, via $V_{EB}'$, the base. The output signal will be developed across $R_L$, one side of which is connected to the collector and the other side, via $V_{CC}$, to the base (note, the d.c. supplies may be considered short circuit as far as a.c. signals are concerned, see section 5.1). The base is therefore common to the input and output circuits and, accordingly, the transistor is said to be operating in the *common base* mode.

The intention is that the input signal should vary the emitter current and consequently the collector current. The forward biasing of the emitter-base junction by $V_{EB}'$ is obviously essential to this end since, without it, the positive half-cycle of the input signal would reverse bias this junction, cutting off the emitter current. With the presence of $V_{EB}'$ however, the positive half-cycle merely reduces the amount of forward bias and lowers the emitter current accordingly. On the other hand the negative half-cycle of the input signal increases the forward bias and therefore increases the emitter current. The collector current is therefore varied by both positive and negative half-cycles of the input signal.

The resistance which the forward biased emitter-base junction presents to the a.c. input signal (i.e. the input resistance $r_{in}$) is very low, say 50 Ω. If the input signal $v_i$, has an amplitude of say 10 mV, the signal component of the emitter current $i_e$, will be:

$$i_e = \frac{v_i}{r_{in}}$$
$$= \frac{10 \times 10^{-3}}{50} \text{ A}$$
$$= 0 \cdot 2 \text{ mA}$$

Since the collector current is typically more than 98% of the emitter current, the resulting signal component of the collector current $i_c$, is approximately 0·2 mA also. This collector signal current will result in variations in the p.d. across the load resistor $R_L$. These variations are the output signal $v_o$, and its amplitude is given by:

$$v_o = i_c R_L$$

If we set the resistance of $R_L$ at 2 kΩ, then:

$$v_o = 0·2 \times 10^{-3} \times 2 \times 10^3 \text{ V}$$
$$= 0·4 \text{ V}$$

The circuit has therefore increased the amplitude of the signal from 10 mV to 0·4 V, that is, the output is 40 times greater than the input.

## 4.4 Common base input characteristic

The relationship between the emitter-base p.d. and the resulting emitter current, for a silicon *npn* transistor, is shown in Fig. 4.7. Since the base and emitter form a *p-n* junction, the emitter current rises in an essentially exponential manner. With a silicon *n-p* junction, a forward p.d. of approximately 0·6 V is required before significant conduction is obtained.

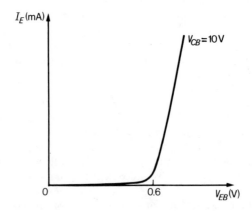

Fig 4.7 Input characteristic for a silicon transistor in the common-base mode. Note. If the transistor being considered is an *npn* type, the emitter-base junction is forward biased when the emitter is negative with respect to the base (see Fig 4.3). Strictly, therefore, the $V_{EB}$ axis should be scaled negatively.

A collector-base p.d. of 10 V is shown against the characteristic. This is the potential at which the collector was held fixed whilst the measurements for the curve were obtained. If the characteristic was replotted for a higher $V_{CB}$, the curve would move marginally to the left. This indicates that the emitter-base p.d., for any particular value of emitter current, decreases as $V_{CB}$ increases. This reduction in $V_{EB}$ is not however significant, and can usually be ignored.

## The Bipolar Transistor

It was stated in section 2.6, that the p.d. across a forward biased junction for any particular current level, tends to decrease by about 2 mV for every degree centigrade rise in junction temperature. This is also the case for this characteristic.

The input characteristic for a germanium transistor is similar in shape to that of the silicon transistor, but significant conduction occurs at a forward p.d. of approximately 0·2 V. The dependence of the characteristic on collector potential and temperature is similar to the silicon device.

Section 4.3 referred to the input resistance of the transistor, i.e. the resistance encountered by a signal applied to the forward biased emitter-base junction. This is the *a.c. input resistance*. The magnitude of this resistance can be determined by reference to the input characteristic which is redrawn in Fig. 4.8. The forward bias p.d. $V_{EB}'$ causes the direct component of emitter current $I_E'$. The input signal $v_i$, which is added to $V_{EB}'$, produces the emitter current $i_e$ (i.e. point $P'$ corresponds to $P$ and so on). The a.c. input resistance in the common base mode is therefore the ratio of these two signal quantities.

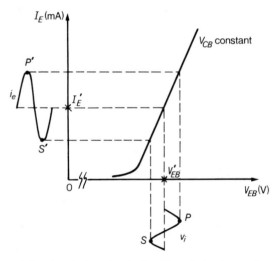

Fig 4.8 Common-base input characteristic (the $V_{EB}$ scale has been enlarged for clarity). The input signal $v_i$ produces the signal component of the emitter current $i_e$.

i.e.
$$r_{in} = \left. \frac{v_i}{i_e} \right| V_{CB} \text{ constant}$$

Since the curve in Fig. 4.8 applies for constant $V_{CB}$, and in a practical amplifier $V_{CB}$ will be changing, this is not strictly accurate. However, it was stated earlier that the effect on the input characteristic of changes in $V_{CB}$ is small, and so it is sufficiently accurate to deduce $r_{in}$ in this way.

Since the input characteristic is non-linear, the value of $r_{in}$ will vary with the point on the curve (i.e. the d.c. bias) about which the signal excursions occur.

The value of $r_{in}$ may alternatively be found for a particular point on the curve, by evaluating the slope (actually the inverse of the slope, since it is the horizontal component divided by the vertical component) of a tangent drawn at the point. Consider the diagram in Fig. 4.9.

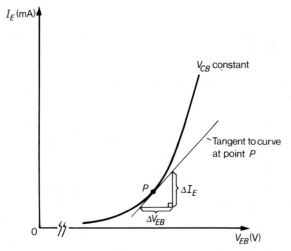

Fig 4.9 Common base input characteristic (the $V_{EB}$ scale has been enlarged for clarity).

The a.c. input resistance at point $P$ can be found using: $r_{in} = \dfrac{\Delta V_{EB}}{\Delta I_E}$.

The a.c. input resistance at point $P$, is:

$$r_{in} = \dfrac{\Delta V_{EB}}{\Delta I_E} \bigg| \; V_{CB} \text{ constant}$$

This approach may be used if the signal $v_i$ in Fig. 4.8 is too small to allow reasonable accuracy to be obtained using that method. A typical range of values for $r_{in}$ in common base is from about 10Ω to 300Ω.

### 4.5 Common base output characteristic

The relationship between the collector current $I_C$ and the collector-base p.d. $V_{CB}$, for various values of emitter current is shown for a silicon *npn* transistor in Fig. 4.10. Similar characteristics would apply for a germanium transistor.

These characteristics clearly show how little the collector current is affected by the collector-base p.d., significant change in collector current occurring only when the emitter current is changed.

# The Bipolar Transistor

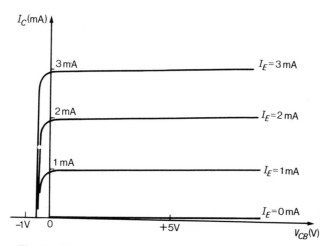

Fig 4.10 Common base output characteristics for an *npn* transistor.

It can be seen from the characteristics that the collector current does not fall to zero when the p.d. between the collector and emitter is zero. This is because the collector-base barrier potential still exists when $V_{CB}$ is zero. Remember that the charge carriers injected from the emitter into the base, are swept across into the collector by this barrier potential. The collector will therefore continue to pick up the injected charge carriers until the barrier potential is overcome by forward biasing the collector-base junction.

The output characteristics can be obtained from measurements taken from a circuit of the type shown in Fig. 4.11.

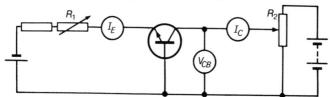

Fig 4.11 Circuit from which measurements can be taken to plot the common base output characteristics of an *npn* transistor.

The procedure is as follows:

(i) The desired value of $I_E$ is set by adjusting $R_1$
(ii) $V_{CB}$ is increased in steps by adjusting $R_2$, and the values of $V_{CB}$ and $I_C$ are recorded at each step
(iii) $I_E$ is set to a new value and (ii) is repeated

If these results are plotted, a family of curves of the type shown in Fig. 4.10 will be obtained. A similar set of characteristics can be obtained for a *pnp* transistor. However, in order to make the necessary measurements, the supply polarities shown in Fig. 4.11 would obviously have to be reversed.

When using a transistor as a common base amplifier, it is important to know the relationship between the input and output currents, that is, the emitter and collector signal currents. Figure 4.12 shows the collector signal current $i_c$, that results from an emitter signal current $i_e$, for a constant collector potential.

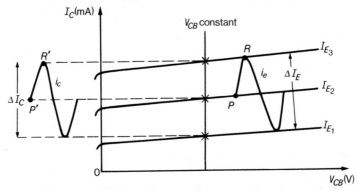

Fig 4.12 Collector signal current $i_c$ resulting from the emitter signal current $i_e$, when $V_{CB}$ is held constant. Note: The upward slope of these characteristics has been exaggerated for clarity.

The signal current $i_e$ is added to the direct component of emitter current $I_{E_2}$. When $i_e$ is zero (point P), the corresponding collector current is the value which applies at the intersection of the $I_{E_2}$ characteristic and the line of constant collector potential. This provides point $P'$ on the $i_c$ waveform. When $i_e$ is at its peak value (point R), it results in the peak value $R'$ on the collector current waveform, and so on. The ratio of the output signal current to the input signal current, is known as the common base current gain $\alpha$, of the transistor.

i.e.
$$\alpha = \frac{i_c}{i_e} \bigg|  V_{CB} \text{ constant}$$

Defining the current gain with $V_{CB}$ constant implies that the collector voltage does not vary, but in an amplifier (Fig. 4.6), the signal current $i_c$ will produce a signal voltage across the load resistor $R_L$. The collector-emitter p.d. is not therefore constant. It will be constant however, if the load resistance $R_L$ is zero (i.e. the output terminals short circuited for a.c. signals) and for this reason $\alpha$ is referred to as the *short circuit common base current gain*.

In Fig. 4.12, $i_e$ and $i_c$ are alternatively represented as incremental changes in $I_E$ and $I_C$ respectively. Therefore:

$$\alpha = \frac{i_c}{i_e} = \frac{\Delta I_C}{\Delta I_E} \bigg| V_{CB} \text{ constant}$$

## The Bipolar Transistor

The value of $\alpha$ for typical low power transistors lies in the range say, 0·95 to 0·997.

It was stated at the beginning of this section, that the collector current is not significantly affected by changes in the collector potential. However, the output characteristics do have a slight upward slope indicating that there is a slight increase in collector current when the collector potential is raised. What this means in effect, is that the resistance which the collector presents to changes in collector potential is large, because the resulting change in collector current is small. This resistance is known as the *a.c. output resistance* in common base. It may be defined as the ratio of a small change in collector-base p.d. $\Delta V_{CB}$, and the corresponding change in collector current $\Delta I_C$, the base current being held constant.

i.e. $$r_{out} = \frac{\Delta V_{CB}}{\Delta I_C}\bigg|_{I_E \text{ constant}}$$

This is illustrated by the enlarged portion of one of the output characteristics shown in Fig. 4.13. A typical range of values for $r_{out}$ in common base (for low power transistors with quiescent collector currents of say 2 mA) is between 50 kΩ and 500 kΩ.

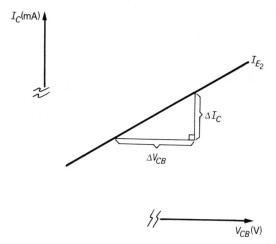

Fig 4.13 Much enlarged portion of a common-base output characteristic. $r_{out} = \dfrac{\Delta V_{CB}}{\Delta I_C}$.

An alternative way to present some of the information contained in the output characteristics, is as a graph of output current against input current. This characteristic, known as the *transfer characteristic*, is drawn for a particular value of $V_{CB}$ as in Fig. 4.14.

For an ideal transistor this characteristic would be a perfectly straight line from the origin. From the diagram, it can be seen that the slope of the transfer characteristic is equal to α.

i.e.
$$\alpha = \frac{\Delta I_C}{\Delta I_E}\bigg|\, V_{CB} \text{ constant}$$

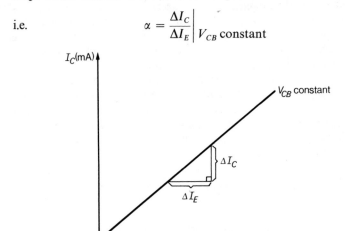

Fig 4.14 Common base transfer characteristic. $\alpha = \dfrac{\Delta I_C}{\Delta I_E}$.

## 4.6 Common emitter mode of connection

Consider the circuit of Fig. 4.15 in which the d.c. supplies to the transistor are arranged differently to those shown in Fig. 4.4. The base-emitter junction is forward biased by $V_{BE}$. The collector supply $V_{CE}$ is applied between the collector and emitter as opposed to collector and base. $V_{CE}$ is normally much larger than $V_{BE}$ and therefore, with respect to the emitter, the collector is more positive than the base. The collector-

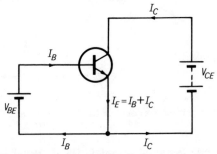

Fig 4.15 Illustrating the d.c. requirements for the common emitter connection of an *npn* transistor.

base junction is therefore reverse biased by an amount equal to the difference between $V_{CE}$ and $V_{BE}$. Since the base-emitter junction is for-

# The Bipolar Transistor

ward biased, the conditions are correct for transistor action to take place. The amount of forward bias that $V_{BE}$ provides, will dictate the rate at which the emitter injects electrons into the base, and hence the emitter and collector currents. However, it is clear from Fig. 4.15 that $V_{BE}$ does not supply the emitter current. It only supplies the base current $I_B$. If $V_{BE}$ is increased, $I_B$ will increase, and so also will $I_E$ and $I_C$. In effect therefore, in this mode of connection, the base current $I_B$ controls the collector current $I_C$.

When connected as an amplifier (as in Fig. 4.16) the input signal is applied between the base, and via $V_{BE}'$, the emitter. The output signal is developed across $R_L$, one side of which is connected to the collector and the other side via $V_{CC}$ to the emitter. The emitter is therefore common to the input and output circuits as far as a.c. signals are concerned and, accordingly, the transistor is described as being in the *common emitter* mode. Note the d.c. supplies may be considered to be

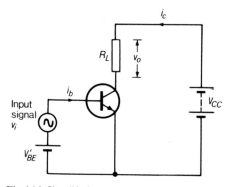

Fig 4.16 Simplified common emitter amplifier.

short circuit as far as a.c. signals are concerned (see section 5.1). A small input signal current $i_b$ results in a much larger output signal current $i_c$ and so there is considerable current amplification. The output signal voltage $v_o$, developed as a result of $i_c$ in the resistor $R_L$, can therefore, be many times greater than $v_i$. This circuit is not a practical circuit but it shows the necessary d.c. conditions. Practical common emitter amplifier circuits are discussed in Chapter 5.

## 4.7 Common emitter input characteristics

The relationship between the base-emitter p.d. and the resulting base current for a silicon transistor is shown in Fig. 4.17. Once again, because we are considering a silicon junction, a forward p.d. of approximately 0·6 V is required before significant conduction is obtained. The precise position of the curve will alter marginally with changes in temperature and collector potential, indicating that the value of $V_{BE}$ required for a particular base current is affected by these two variables.

The trend of these variations is similar to those outlined for the common base input characteristic described in section 4.4. A similar curve will be obtained for a germanium transistor except that significant conduction is obtained for a forward base-emitter p.d. of approximately 0·2 V.

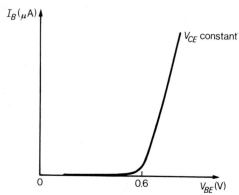

Fig 4.17 Input characteristic for a silicon transistor in common emitter mode.

The *a.c. input resistance* of the transistor in common emitter may be found using an approach similar to that adopted for common base in section 4.4.

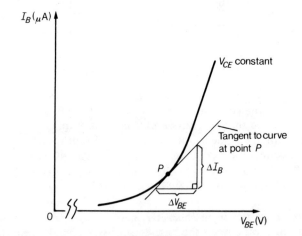

Fig 4.18 Common-emitter input characteristic (the $V_{BE}$ scale has been enlarged for clarity). The a.c. input resistance may be found at point $P$ using: $r_{in} = \dfrac{\Delta V_{BE}}{\Delta I_B}$.

The a.c. input resistance at point $P$ on the characteristic drawn in Fig. 4.18 is given by:

# The Bipolar Transistor

$$r_{in} = \left.\frac{\Delta V_{BE}}{\Delta I_B}\right|_{V_{CE} \text{ constant}}$$

A typical range of values for $r_{in}$ in common emitter is from about 1 kΩ to 5 kΩ.

A circuit suitable for obtaining measurements from which to plot the input characteristic is shown in Fig. 4.19. Since base currents in the

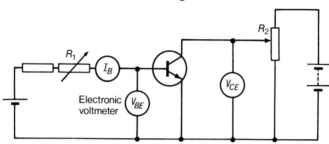

Fig 4.19 Circuit from which measurements can be taken to plot the common emitter input characteristic.

region of tens of microamperes are to be expected, care must be taken in the choice of instrument used to measure the base-emitter p.d. An ordinary moving coil voltmeter may require 50 µA or so for full-scale deflection (f.s.d.). In the circuit of Fig. 4.19, such voltmeter currents would result in the microammeter indicating a value much in excess of the true base current. To avoid this problem, the base-emitter p.d. is measured by an electronic voltmeter. A typical electronic voltmeter will require an f.s.d. current of much less than 1 µA in this situation. The procedure to be adopted with the circuit of Fig. 4.19 is as follows:

(i) Set the value of $V_{CE}$
(ii) Increase $I_B$ in steps by adjusting $R_1$ (ensuring at each step that $V_{CE}$ is unchanged), recording $V_{BE}$ and $I_B$ at each step.
(Note, after establishing the general shape of the characteristic in this manner, it may be necessary to make some additional measurements in order to accurately plot the 'knee' of the characteristic).

## 4.8 Common emitter output characteristics

The output characteristics for an *npn* transistor connected in the common emitter mode are drawn in Fig. 4.20 (similar output characteristics would be obtained for a *pnp* transistor).

At collector potentials above 0·5 V or so, the collector potential has little effect on the collector current. Notice that for these characteristics, unlike those for common base, the collector current is reduced to practically zero when $V_{CE}$ is reduced to zero. This is because when the

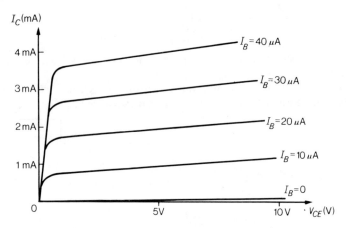

Fig 4.20 Common emitter output characteristics.

collector potential is reduced below the base potential, the collector-base junction becomes forward biased. This forward bias reduces the collector-base barrier potential and normal transistor action ceases. Instead, the collector carries a small proportion of the base current as a forward bias current. This situation is illustrated in Fig. 4.21 by considering, for this purpose, the transistor as two diodes.

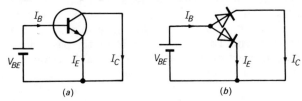

Fig 4.21 Diagram (b) represents the transistor as two diodes in order to illustrate the situation when $V_{CE} = 0$.

Figure 4.22 shows the collector signal current $i_c$, that results from the base signal current $i_b$, the collector potential being held constant. The *short circuit current gain* in common emitter $\beta$* (that is the current gain with $R_L$ zero), is defined as follows:

$$\beta = \frac{i_c}{i_b}\bigg|\ V_{CE}\ \text{constant}$$

Since $i_c$ and $i_b$ are incremental changes in $I_C$ and $I_B$ respectively, then alternatively:

---

* The short circuit current gain in common emitter is often quoted as $h_{fe}$. This symbol comes from the transistor hybrid or 'h'-parameters which are beyond the scope of this book. It has been mentioned because it is commonly given in manufacturer's data.

# The Bipolar Transistor

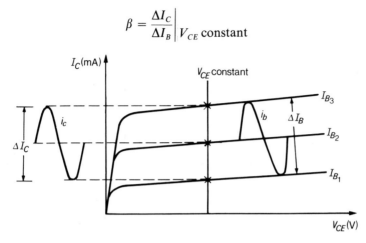

Fig 4.22 Collector signal current $i_c$ resulting from base signal current $i_b$ when $V_{CE}$ is held constant.

The value of $\beta$ for typical low power transistors lies between approximately 40 and 450.

It should be recognised that $\beta$ is frequently used to indicate the relationship between the direct collector and base currents, as well as the signal collector and base currents. That is, sometimes $\beta = \dfrac{I_C}{I_B}$. This is because, for practical purposes, the same numerical value for $\beta$ may often be used without significant error, whether considering signal or direct currents in circuit calculations.

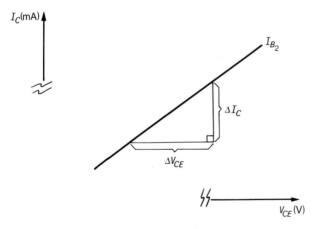

Fig 4.23 Much enlarged portion of common-emitter output characteristic. $r_{out} = \dfrac{\Delta V_{CE}}{\Delta I_C}$.

The *a.c. output resistance* for the transistor in common emitter is found, as it was for common base, by dividing an incremental change in collector-emitter p.d. $\Delta V_{CE}$, by the resulting incremental change in collector current $\Delta I_C$, the base current being held constant. This is illustrated by the enlarged portion of an output characteristic shown in Fig. 4.23. The a.c. output resistance is therefore:

$$r_{out} = \frac{\Delta V_{CE}}{\Delta I_C}\bigg|\, I_B \text{ constant}$$

Typical values for $r_{out}$ for low power transistors in common emitter (with quiescent collector currents of say 2 mA) lie in the range from say 10 kΩ to 100 kΩ.

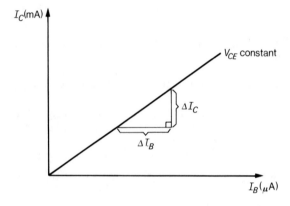

Fig 4.24 Common emitter transfer characteristic. $\beta = \dfrac{\Delta I_C}{\Delta I_B}$.

The *transfer characteristic* for the transistor in common emitter is shown in Fig. 4.24. For the ideal transistor, this characteristic would be a perfectly straight line from the origin. It can be seen from Fig. 4.24 that the gradient of the transfer characteristic is equal to $\beta$, i.e.

$$\beta = \frac{\Delta I_C}{\Delta I_B}\bigg|\, V_{CE} \text{ constant}$$

### 4.9 Relationship between $\beta$ and $\alpha$

For any transistor, there is a definite relationship between its short circuit current gain in common emitter and its short circuit current gain in common base. This relationship can be derived as follows:

$$I_E = I_C + I_B$$
$$\therefore \quad \Delta I_E = \Delta I_C + \Delta I_B$$

*The Bipolar Transistor*

Since an incremental change in a current can be regarded as a signal component, this can be written as:

$$i_e = i_c + i_b$$

Dividing both sides by $i_c$:

$$\frac{i_e}{i_c} = \frac{i_c}{i_c} + \frac{i_b}{i_c}$$

But $\alpha = \dfrac{i_c}{i_e}$, and $\beta = \dfrac{i_c}{i_b}$, hence substituting these relationships in the above equation:

$$\frac{1}{\alpha} = 1 + \frac{1}{\beta}$$

$\therefore$
$$\frac{1}{\beta} = \frac{1}{\alpha} - 1$$

$$= \frac{1 - \alpha}{\alpha}$$

$\therefore$
$$\beta = \frac{\alpha}{1 - \alpha}$$

Putting $\alpha = 0.99$

$$\beta = \frac{0.99}{1 - 0.99}$$

$$= 99$$

Similarly, it can be shown that

$$\alpha = \frac{\beta}{1 + \beta}$$

**Problems**
(Answers at end of book)

1  Describe the basic principle of operation of the bipolar transistor, explaining in particular detail:
   (i) why the majority of the charge carriers that are injected into the base from the emitter, are picked up by the collector

and

   (ii) why the collector current tends to be almost unaffected by the magnitude of the collector potential.

2  The figures given in the table below were obtained for a silicon *npn* transistor connected in the common emitter mode. Plot the input characteristic for the transistor and hence determine its input re-

sistance for an input signal of 20 mV peak-to-peak, added to a quiescent base-emitter p.d. of 0·625 V.

| $V_{BE}$ (V) | 0·55 | 0·575 | 0·6 | 0·625 | 0·65 |
|---|---|---|---|---|---|
| $I_B$ (μA) | 0·5 | 1·5 | 5·0 | 15 | 25 |

3  The table below shows the results that were obtained from measurements made on a silicon *npn* transistor connected in the common emitter mode. Plot the static output characteristics for the transistor. Using quiescent values of base current and collector potential of 24 μA and 5 V respectively, determine (i) $h_{fe}$ and (ii) $r_{out}$.

|  | $I_c$(mA) | | |
|---|---|---|---|
| $V_{CE}$ | 1 V | 6 V | 10 V |
| $I_B = 8$ μA | 1 | 1·05 | 1·1 |
| $I_B = 16$ μA | 2·05 | 2·12 | 2·18 |
| $I_B = 24$ μA | 3·1 | 3·18 | 3·26 |
| $I_B = 32$ μA | 4·16 | 4·29 | 4·38 |
| $I_B = 40$ μA | 5·3 | 5·45 | 5·55 |

4  For a bipolar transistor, deduce an expression for $\alpha$ in terms of $\beta$ (starting from the relationship: $i_e = i_c + i_b$). Hence evaluate $\alpha$ if $\beta$ is (i) 40 and (ii) 400.

# Chapter 5
# The Transistor Amplifier

## 5.1 Basic common emitter amplifier

A basic transistor amplifier circuit is represented in Fig. 5.1. The input signal $v_b$ is applied between the base and emitter of the transistor. This produces a signal current in the base which is amplified by transistor action to produce a corresponding but substantially larger collector signal current. This collector signal current in the load $R_L$ produces the output signal $v_c$. The magnitude of the output signal may be many times greater than the input signal, i.e. a substantial voltage amplification can be achieved.

The input signal is applied between the base and emitter, and the output signal is developed between the collector and the emitter. The emitter is therefore common to the input and output circuits and hence this is called a common emitter amplifier.

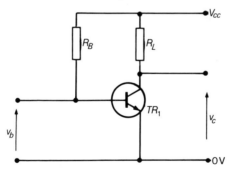

Fig 5.1 Basic common emitter amplifier.

A common emitter amplifier is an inverting amplifier, i.e. the output signal $v_c$ is inverted relative to the input signal $v_b$. The reason for this becomes clear when the action of $TR_1$, which in this case is an *npn* transistor, is considered. A positive increase in the voltage at the base will increase the base current and hence increase the collector current. This will result in an increased p.d. across $R_L$ and therefore a reduction in the collector voltage. So, a positive increase at the base, causes a decrease in the collector voltage (i.e. the collector becomes less positive). Similarly, if the base voltage is reduced, the base current, and hence the collector current will be reduced. The collector voltage will therefore rise. Clearly then, the output signal is inverted relative to the input signal.

Now it has been established that a change in p.d. across $R_L$ results in a change in the collector voltage. These two changes are, of course, the same in magnitude. In fact, the signal p.d. across the load resistor and that across the transistor are one and the same. This becomes apparent when a simple a.c. (i.e. signal) equivalent circuit is drawn for the amplifier as in Fig. 5.2(b). The d.c. requirements are ignored and only signal current paths are considered. Note that the $V_{cc}$ and $0V$ supply lines have been joined together in this diagram. This is because, provided the supply has negligible internal resistance, no signal p.d. is developed across the d.c. supply terminals. In other words they are at the same signal potential and, as far as a.c. conditions are concerned, they can be assumed to be joined together. In fact, in practical amplifier circuits, the supply lines are commonly linked by a capacitor ($C_1$ in Fig. 5.2(a)), the reactance of which is negligible at signal frequencies. $R_L$ is now seen to be in parallel with the transistor. The signal between collector and emitter and that across $R_L$ are now clearly one and the same.

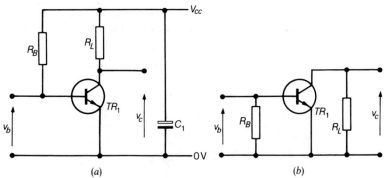

Fig 5.2 (a) Basic common emitter amplifier.
(b) Effective a.c. equivalent circuit of the amplifier in Fig. 5.2(a) with $C_1$ considered to be a short circuit to a.c. signals.

## 5.2 Operating point

The *operating point*, or *working point*, of the transistor specifies its d.c. operating conditions. These are the conditions when no signal is applied and, for a given transistor, are set by the supply voltage and the resistances $R_B$ and $R_L$. The common emitter output characteristics for the transistor in Fig. 5.3(a) are shown in Fig. 5.3(b). The operating point is marked on the characteristics as point $Q$. The values of collector voltage and collector current which apply at this point are called the no-signal, or *quiescent* values. From Fig. 5.3(b), the quiescent col-

## The Transistor Amplifier

lector voltage is 5 V and the corresponding quiescent collector current is 2 mA. Assuming the amplifier is to be used to amplify a sine wave signal (i.e. a signal with equal positive and negative excursions) this operating point will permit the maximum undistorted signal swing at the collector. The collector voltage can, if necessary, rise by 5 V to the supply voltage of 10 V during the positive excursion of the output signal. Similarly it can fall by 5 V to 0 V during the negative excursion.* These are obviously the largest possible changes in output voltage. Attempts to increase the amplitude of the output signal any further will result in distortion, i.e. clipping of the output signal. The largest undistorted output signal that can be obtained is therefore about 10 V p–p (where p–p means peak to peak value).

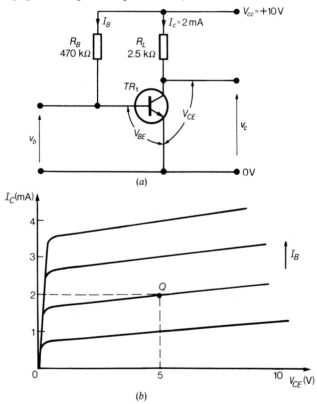

Fig 5.3 (a) Basic common emitter amplifier.
(b) Quiescent Operating Point.

---

* In practice, the collector voltage cannot be reduced to zero because there will be a small saturation p.d. between collector and emitter. The maximum undistorted output that may be obtained is therefore a little less than 10 V p–p.

If $V_{CE}$ is the quiescent p.d. between collector and emitter and $V_{RL}$ is the p.d. across the load, then, since the sum of these two must equal the supply voltage:

$$V_{CE} + V_{RL} = V_{CC}$$

$$\therefore V_{RL} = V_{CC} - V_{CE}$$
$$= 10 - 5 \text{ V}$$
$$= 5 \text{ V}$$

The collector current $I_C$ is 2 mA, and hence the resistance $R_L$ is:

$$R_L = \frac{V_{RL}}{I_C}$$
$$= \frac{5}{2 \times 10^{-3}} \Omega$$
$$= 2 \cdot 5 \text{ k}\Omega$$

In practice, the nearest preferred value,† of 2·7 kΩ would be used but it is convenient to leave this resistance as 2·5 kΩ for the moment.

The desired collector current of 2 mA is established by choosing the appropriate resistance for $R_B$. This resistor determines the quiescent base current of the transistor. Assume the quiescent p.d. between the base and emitter is 0·6 V (typical for a silicon transistor).

The p.d. across $R_B$ is then:

$$V_{RB} = V_{CC} - V_{BE}$$
$$= 10 - 0 \cdot 6 \text{ V}$$
$$= 9 \cdot 4 \text{ V}$$

If the current gain $\beta$* is 100, the base current $I_B$ is:

$$I_B = \frac{I_C}{\beta}$$
$$= \frac{2 \times 10^{-3}}{100} \text{ A}$$
$$= 20 \text{ } \mu\text{A}$$

---

† It is clearly impractical, and indeed unnecessary, to manufacture resistors with every conceivable value of resistance. In practice, resistors are manufactured to a number of standardised nominal resistance values called *preferred values*. In most circuits, the precise resistance values are not critical and so resistors with the nearest preferred values to those calculated for the circuit are selected.

* $\beta$ is taken to represent both d.c. and a.c. current gain which are assumed to be equal (see section 4.8).

# The Transistor Amplifier

The base resistor $R_B$ is then:

$$R_B = \frac{V_{RB}}{I_B}$$

$$= \frac{9 \cdot 4}{20 \times 10^{-6}} \Omega$$

$$= 470 \text{ k}\Omega$$

This would appear to be sufficient to establish the specified operating point. However, this simple biasing arrangement using only $R_P$, although functional, leaves the collector current very dependent on both temperature and transistor parameters. In practice, a precise and repeatable operating point cannot be achieved by this method. This circuit is however adequate for the moment and the problem of stabilising the operating point is considered further in section 5.10.

## 5.3 Load line

So far, only the quiescent operating point has been considered. The operation of the circuit as an amplifier will now be examined graphically by constructing what is called a *load line*. A load line shows, for a particular load resistance, the values of collector voltage to which the transistor is restricted for given values of collector current. It can be established by considering particular conditions, such as those which apply when the collector current is zero and those which apply when the collector voltage is zero. Now, the collector current could be reduced to zero by applying a sufficiently large negative input signal excursion at the base. If this were to occur there would be no p.d. across $R_L$ and the p.d. between collector and emitter would be equal to

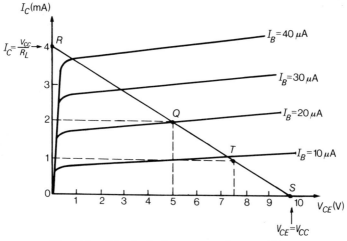

Fig 5.4 Load line for basic common emitter amplifier.

the supply voltage. This condition is plotted as point $S$ in Fig. 5.4. (i.e. $V_{CE} = 10$ V, $I_C = 0$).

Similarly, a large positive input signal excursion could reduce $V_{CE}$ to virtually zero. The full supply voltage would then appear across $R_L$. The corresponding collector current would be:

$$I_C = \frac{V_{CC}}{R_L}$$
$$= \frac{10}{2 \cdot 5 \times 10^{-3}} \text{ A}$$
$$= 4 \text{ mA}$$

This gives point $R$ (i.e. $V_{CE} = 0$, $I_C = 4$ mA).

The conditions for other values of collector current (brought about by other input signal levels of course) may be plotted in a similar fashion. For example, if the collector current is 1 mA, then:

$$V_{RL} = I_C R_L$$
$$= 1 \times 10^{-3} \times 2 \cdot 5 \times 10^3 \text{ V}$$
$$= 2 \cdot 5 \text{ V}$$
$$\therefore \quad V_{CE} = V_{CC} - V_{RL}$$
$$= 10 - 2 \cdot 5 \text{ V}$$
$$= 7 \cdot 5 \text{ V}$$

This is plotted as point $T$, (i.e. $V_{CE} = 7 \cdot 5$ V, $I_C = 1$ mA).

By repeating this operation for many values of collector current, the load line $RS$, can be established. In practice, since the load line is a straight line, only two points are required in order to position it. It may therefore be drawn using just the co-ordinates of the points $R$ and $S$ which are easily determined, as shown above. The load line may be used to show how the collector current and collector voltage are related to the base current. In Fig. 5.5, the signal current in the base $i_b$ is plotted on the output characteristics. At point $L$, this signal current is zero. The base current is therefore the quiescent value of 20 $\mu$A, and the corresponding collector current is 2 mA (point $L'$). This value is determined by referring the $I_B = 20$ $\mu$A curve to the load line, i.e. it is the value of the collector current at the intercept of these two. At $M$, the signal current is 10 $\mu$A, which, added to the quiescent base current, gives a net base current of 30 $\mu$A. The point of intersection of the 30 $\mu$A curve and the load line gives a collector signal current of 2·85 mA (point $M'$). By repeating this operation for many signal current values, the complete collector signal current waveform may be constructed against the $I_C$ axis, as shown. In a similar manner, the corresponding collector signal voltage may be constructed against the $V_{CE}$ axis by relating $M$ to $M''$ and so on.

# The Transistor Amplifier

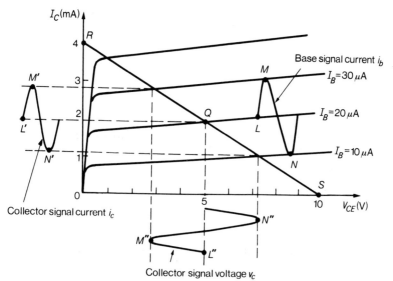

Fig 5.5 Relationship between the collector signal current, the collector signal voltage and the base signal current demonstrated by using a load line.

## 5.4 Current amplification

The short circuit current gain $\beta$ for the transistor can be evaluated at the operating point $Q$ by the method indicated in section 4.8. In this case $\beta$ is 100. This is not the current gain of the amplifier however. The *signal current gain* of the amplifier $A_i$, is the ratio of the output and input signal currents,

i.e. $$A_i = \frac{i_{out}}{i_{in}}$$

The output signal current $i_{out}$ is equal to the collector signal current $I_c$ deduced using the load line in Fig. 5.5. The input signal current $i_{in}$ is not however equal to $i_b$. This is because the signal current delivered to the amplifier is shared between the bias resistor $R_B$ and the transistor base. The simple a.c. equivalent circuit which is redrawn in Fig. 5.6 shows how this input signal current is divided.

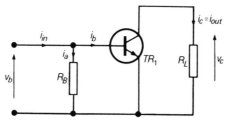

Fig 5.6 Simple a.c. equivalent circuit showing how the input current is shared between the bias resistor $R_B$ and the transistor base.

Clearly $i_{in} = i_a + i_b$, where $i_a = \dfrac{v_b}{R_B}$

and $i_b = \dfrac{v_b}{r_{in}}$

(where $r_{in}$ is the a.c. input resistance of the transistor)

In the basic amplifier of Fig. 5.3(a), $R_B$ will be very much greater than $r_{in}$ and therefore $i_a \ll i_b$. In this case therefore, $i_a$ may be neglected and the input current may be assumed to be equal to $i_b$. The current gain can then be found by reference to Fig. 5.5.

i.e.
$$A_i = \frac{i_{out}}{i_{in}}$$

$$= \frac{i_c}{i_b} \quad \text{if } R_B \gg r_{in}$$

Taking the peak-to-peak signal current values from the diagram

$$A_i = \frac{1 \cdot 7 \times 10^{-3}}{20 \times 10^{-6}}$$

$$= 85$$

If $R_B$ is not many times greater than $r_{in}$, then it can significantly reduce the current gain of the amplifier. By way of example, consider the very unlikely case where $R_B$ is so small that it is actually equal to $r_{in}$. The input signal current would divide equally between $R_B$ and the transistor, and therefore only half of it would enter the transistor base to be amplified by transistor action. The base signal current $i_b$ would then be $\dfrac{i_{in}}{2}$ and the current gain of the amplifier would be halved.

### 5.5 Voltage amplification

The voltage amplification, or *voltage gain* $A_v$, of the amplifier in Fig. 5.3(a) is the ratio of the output signal voltage $v_c$, to the input signal voltage $v_b$

$$A_v = \frac{v_c}{v_b}$$

The relationship between the output signal voltage $v_c$ and the signal current at the base $i_b$ has already been illustrated in Fig. 5.5. The corresponding input signal voltage $v_b$ may be found by reference to the input characteristic of the transistor. This is illustrated in Fig. 5.7.*

---

* Input characteristics are normally plotted with $V_{CE}$ constant. In an amplifier, $V_{CE}$ is obviously changing because an output signal is being developed. However, changes of

# The Transistor Amplifier

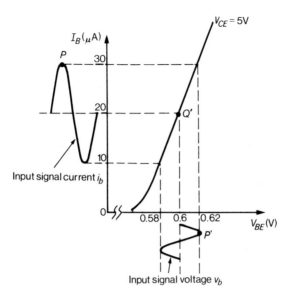

Fig 5.7 Input characteristic for $TR_1$ showing the relationship between $v_b$ and $i_b$.

Point $Q'$ corresponds to the quiescent operating point $Q$ on the output characteristics, i.e. $I_B = 20$ μA. The input signal current waveform is constructed against the $I_B$ axis and the corresponding values of $V_{BE}$ are found by reference to the curve (i.e. $P$ corresponds to $P'$ etc.). From this, the input signal voltage is found to be 40 mV p–p.

The output signal voltage taken from Fig. 5.5 is 4·25 V p–p. The magnitude of the voltage gain is thus:

$$|A_v| = \left|\frac{v_c}{v_b}\right|$$

where $|A_v|$ represents the magnitude (or modulus) of the voltage gain

$$= \frac{4\cdot25}{40 \times 10^{-3}}$$

$$= 106$$

To conclude the graphical analysis of this amplifier, we shall show that

---

$V_{CE}$ have only a small effect and the characteristic that applies for the quiescent value of $V_{CE}$ may be used. This characteristic is drawn with a suppressed zero for clarity. The curve is also idealised to some extent. In practice, the input characteristic will be non-linear and the input current will not be a perfect replica of the input signal voltage. This non-linearity may be ignored if the input signal swing is small. Furthermore, if the amplifier is driven from a high resistance source, this non-linearity (which represents changes of $r_{in}$ with instantaneous signal magnitude) is swamped by the large source resistance.

the output resistance of the transistor has influenced the magnitudes of the current gain and voltage gain that have been achieved. That is, in the amplifier in Fig. 5.3(a) the effect of the output resistance of the transistor has been to reduce $A_i$ and $A_v$.

## 5.6 Effect of output resistance on current gain

The effect of the transistor's output resistance $r_{out}$ (remembering $r_{out}$ is found from the slope of the output characteristics) may be illustrated by changing the slope of the output characteristics. Consider the output characteristics shown in Fig. 5.8 which are drawn parallel to the $V_{CE}$ axis. These represent a transistor with an infinite output resistance. Inspection of these characteristics shows that the short circuit current gain $\beta$, at the operating point $Q$, is 100. This was also the case for Fig. 5.5. Making the assumption that $R_B \gg r_{in}$ as before, the signal current gain is found by taking the ratio of $i_c$ and $i_b$ from Fig. 5.8. From the diagram $i_c$ is 2 mA p–p, and the corresponding $i_b$ is 20 μA p–p. The current gain is therefore:

$$A_i = \frac{i_{out}}{i_{in}}$$

and, if $R_B \gg r_{in}$
$$A_i = \frac{i_c}{i_b}$$

$$= \frac{2 \times 10^{-3}}{20 \times 10^{-6}}$$

$$= 100$$

Hence with $r_{out} = \infty$, the current gain $A_i$ is equal to $\beta$. Compare this with the current gain of 85 found in section 5.4. The only difference between these two situations is that, in the example of Fig. 5.4, the transistor has a finite output resistance as the upward slope of the output characteristics indicates. It can be seen then, that the effect of

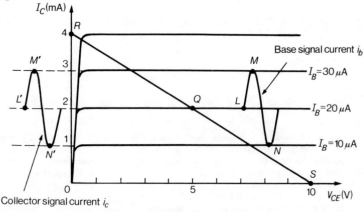

Fig 5.8 Current amplification with $r_{out} = \infty$.

# The Transistor Amplifier

the transistor's output resistance is to reduce the current gain of the amplifier.

From the foregoing it is clear that, when determining the current gain by graphical methods, the effect of the output resistance is automatically taken into account. When deducing the current gain by other means, due consideration must be given to the effect of the output resistance. However, in most practical situations the effect of the output resistance may be ignored provided $r_{out} \gg R_L$.

## 5.7 Effect of output resistance on voltage gain

The output resistance of a transistor also influences the voltage gain of an amplifier. As before, the effect that the output resistance of $TR_1$ will have had on the gain of the amplifier in Fig. 5.3(a), can be assessed by considering an alternative transistor amplifier which is identical in all respects except that the transistor is assumed to have an infinite output resistance. In Fig. 5.9 the relationship between $i_b$ and $v_c$ for such a transistor is illustrated by using the same set of curves as used in Fig. 5.8. The output signal voltage is seen to be 5 V p–p. This applies for the same input signal swing $i_b$ of 20 µA p–p used previously. So, assuming this arises from an input signal voltage $v_b$ of 40 mV p–p, the magnitude of the voltage gain is:

$$|A_v| = \left|\frac{v_c}{v_b}\right|$$
$$= \frac{5}{40 \times 10^{-3}}$$
$$= 125$$

Compare this with the previous figure of 106 found in section 5.5. The ratio of these two values is:

$$\frac{106}{125} = 0{\cdot}85$$

That is, the effect of the transistor's output resistance has been to reduce the voltage gain to 85% of what could be achieved if a transistor with infinite output resistance was possible. The current gain of the amplifier was also seen to be reduced to 85% of $\beta$ when the output resistance was taken into account. It is perhaps not surprising that the current gain and voltage gain are reduced by the same proportion, since the output signal voltage is proportional to the output signal current in $R_L$.

Again, when using graphical techniques, the effect of the transistor's output resistance is automatically taken into account. As before when deducing the voltage gain by other means the effect of the output resistance of the transistor may be neglected provided $r_{out} \gg R_L$.

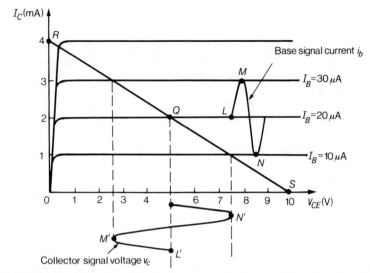

Fig 5.9 Relationship between the signal current at the base and the output signal voltage with $r_{out} = \infty$.

## 5.8 Expressions for current gain and voltage gain

It has been shown that there are two factors which influence the current gain of the amplifier in Fig. 5.3(a). These are the sharing of the input signal current between $R_B$ and the transistor base (section 5.4), and the effect of the transistor's output resistance (section 5.6). However, if we make the two assumptions that $R_B \gg r_{in}$ and $r_{out} \gg R_L$, the current gain $A_i$ may be taken to be equal to $\beta$.

An expression for the voltage gain can be deduced by making the assumption that the input characteristic is linear around the operating point (a perfectly reasonable assumption if the input signal swing is small). The input signal $v_b$ is then given by:

$$v_b = i_b r_{in} \qquad (1)$$

The output signal $v_c$ is produced by the output signal current $i_c$ in the load $R_L$

∴
$$|v_c| = i_c R_L$$

It was shown earlier that a common emitter amplifier is an inverting amplifier. This means that, as the collector current increases, the collector voltage falls.

Therefore,
$$v_c = -i_c R_L \qquad (2)$$

Hence using equations (1) and (2), the voltage gain is:

$$A_v = \frac{v_c}{v_b}$$

# The Transistor Amplifier

$$= \frac{-i_c R_L}{i_b r_{in}}$$

$$\simeq \frac{-\beta R_L}{r_{in}}$$

(since with $r_{out} \gg R_L$, $\frac{i_c}{i_b} \simeq \beta$)

Applying this result to the amplifier in Fig. 5.3(a) with $r_{in} = 2$ k$\Omega$ (deduced from Fig. 5.7)

$$A_v \simeq -\frac{\beta R_L}{r_{in}}$$

$$= -\frac{100 \times 2 \cdot 5 \times 10^3}{2 \times 10^3}$$

$$= -125$$

This of course agrees with the figure obtained graphically with $r_{out} = \infty$.

## 5.9 Power amplification

The power gain is the ratio of the output signal power to the input signal power. It was shown, in section 5.4, that the input signal current is shared between the transistor base and the base bias resistor. The input signal power is similarly shared between these two. Once again, if $R_B \gg r_{in}$, the base bias resistor may be ignored. The input signal power is then:

$$P_{in} = I_b^2 r_{in}$$

The output power is:

$$P_{out} = I_c^2 R_L$$

(where $I_b$ and $I_c$ are the r.m.s. values of input and output signal currents respectively)

The power gain $A_p = \dfrac{P_{out}}{P_{in}}$

$$= \frac{I_c^2 R_L}{I_b^2 r_{in}}$$

Since these currents are expressed as a ratio, they may be replaced by instantaneous quantities

i.e.
$$A_p = \frac{i_c^2}{i_b^2} \frac{R_L}{r_{in}}$$

Putting $\beta = \dfrac{i_c}{i_b}$, i.e. assuming that the output resistance of the transistor is infinite:

$$A_p = \beta^2 \frac{R_L}{r_{in}}$$

Using the previous values, for the circuit in Fig. 5.3(a)

$$A_p = \frac{100^2 \times 2\cdot 5 \times 10^3}{2 \times 10^3}$$

$$= 12\ 500$$

The power gain is commonly quoted in dB's

$$A_p(\text{dB}) = 10 \log_{10} A_p$$
$$= 10 \log_{10} 12\ 500$$
$$= 41 \text{ dB}$$

It is also possible to write the power gain in terms of the voltage gain and current gain.

$$A_p = \beta^2 \frac{R_L}{r_{in}}$$

$$= (\beta)\frac{(\beta R_L)}{r_{in}}$$

$$= A_i | A_v |$$

i.e.     Power gain = current gain × |voltage gain|

It has been shown that the effect of the output resistance of the transistor is to reduce both the current gain and the voltage gain, and so inevitably it will reduce the power gain also.

## 5.10 Stabilisation of the operating point

The simple biasing arrangement using only $R_B$ in Fig. 5.3(a), although functional, is seldom satisfactory in practice. There are two reasons for this:

(i) The actual value of collector current developed by a transistor, for a given value of base current, is temperature dependent. This means that the operating point of the transistor will vary with temperature.

(ii) In order to satisfy mass-production and servicing requirements, it must be possible to insert any transistor of the particular type in use into the circuit and achieve a satisfactory performance. With this simple biasing arrangement, the variations in transistor parameters that are encountered (i.e. transistor spreads) will

## The Transistor Amplifier

result in significant changes of operating point from one transistor to another.

Let us consider the effect of temperature in more detail. The collector to base leakage current for a silicon transistor approximately doubles for every 10°C rise in junction temperature. $\beta$ rises with temperature also. Both of these factors contribute towards an increase of collector current as the operating temperature is raised. This increase of collector current will itself cause a further rise of junction temperature, because it causes increased power dissipation in the transistor. Under certain conditions this effect may become cumulative, i.e. increased collector current, causing increased junction temperature, causing increased collector current, and so on. This effect is then called thermal runaway and may result in permanent damage to the transistor.

When considering transistor spreads, the variation of $\beta$ from one device to another is found to be the most significant factor in the majority of circuits. A particular transistor type, with a quoted typical $\beta$ of 100, may perhaps have a $\beta$ in the range 50 to 500. The effect of this type of variation can be illustrated by reference to the basic amplifier which is redrawn in Fig. 5.10.

$R_B$ was chosen to give a collector current of 2 mA when using a transistor with a $\beta$ of 100 (see section 5.2). If $TR_1$ is replaced by a similar transistor with a $\beta$ of say 180, the collector current will be:

$$I_C = \beta I_B$$
$$= 180 \times 20 \times 10^{-6} \text{ A}$$
$$= 3 \cdot 6 \text{ mA}$$

The base current has been assumed to be the same as before because it is virtually independent of the transistor itself, as shown below.

$$I_B = I_{RB}$$
$$= \frac{V_{RB}}{R_B}$$
$$= \frac{V_{CC} - V_{BE}}{R_B}$$

but $\quad V_{CC} \gg V_{BE}$

$\therefore \quad I_B \simeq \dfrac{V_{CC}}{R_B}$

Now, with a collector current of 3·6 mA, the p.d. across $R_L$ is:

$$V_{RL} = I_C R_L$$
$$= 3 \cdot 6 \times 10^{-3} \times 2 \cdot 5 \times 10^3 \text{ V}$$
$$= 9 \text{ V}$$

The collector voltage is therefore 1 V. At this new operating point only

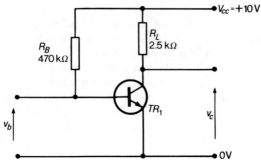

Fig 5.10 Basic common emitter amplifier.

a very small undistorted output signal is possible. This is because negative voltage excursions at the collector, greater than about 0·5 V, will result in this half-cycle of the output signal being clipped.

### 5.11 Self-adjusting bias circuit

The biasing circuit of Fig. 5.11 is commonly used because it automatically adjusts the bias current to compensate the effects of temperature and transistor spreads.

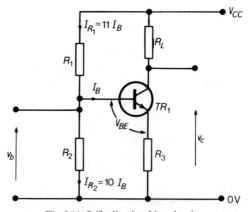

Fig 5.11 Self-adjusting bias circuit.

The potential divider formed by $R_1$ and $R_2$ fixes the quiescent base voltage. This is because the loading effect of the base circuit on the potential divider is insignificant. That is, the values of these two resistors are chosen such that the currents in them are large compared with the base current $I_B$. Commonly $I_{R_2}$ is made about 10 times greater than $I_B$. The base current is of course supplied via $R_1$ and so $I_{R_1}$ is then

# The Transistor Amplifier

11 $I_B$. Since the current in $R_1$ and $R_2$ is large by comparison to the base current, any small change in $I_B$ is not significant and the base voltage can be assumed to be substantially constant. A voltage is established at the emitter due to the emitter current in the resistor $R_3$. The p.d. between the base and emitter $V_{BE}$ is the difference between the fixed base voltage and this emitter voltage.

Now consider the effect of an increase in ambient temperature. This will cause an increase in emitter current and hence an increase in the emitter voltage. Since the base voltage is fixed, this will reduce $V_{BE}$. A reduction in $V_{BE}$ normally causes a reduction in collector current and so, in this situation, its effect will be to offset the original tendency of the collector current to rise.

The circuit also compensates for changes in $\beta$. Assume the circuit was initially designed for a transistor with a $\beta$ of 100. If now, a transistor with a $\beta$ of 150 is selected, the collector and emitter currents might be expected to be 1·5 times greater. However, an increase in emitter current causes an increase in emitter voltage and a reduction of $V_{BE}$ as before. Again, remembering that the base voltage is fixed, this offsets the rise in collector current and a quiescent current close to the desired operating figure is established.

## 5.12 Emitter decoupling

When a signal is applied at the base of the transistor, it causes changes of the collector and emitter currents. With the circuit as shown in Fig. 5.11, a signal voltage is developed across $R_L$ as required, but, because of the presence of $R_3$, a signal is also developed at the emitter. A positive increase in the voltage at the base will increase the emitter current and cause a positive increase in the emitter voltage. The signal developed at the emitter is therefore in phase with that at the base as shown in Fig. 5.12.

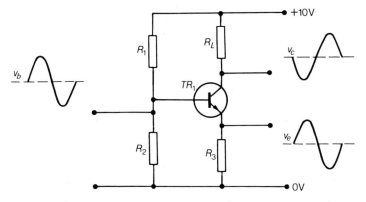

Fig 5.12 Signals are developed at the collector and the emitter when $R_3$ is included. (Note: If $R_L = R_3$ then $|v_c| \simeq |v_e|$ and the circuit is called a phase-splitter.)

Now, if the base and emitter voltages were to increase by the same amount, there would be no net change in $V_{BE}$ and hence no effective input signal. Therefore, the signal at the emitter will always be slightly smaller than the signal at the base, the effective input signal being the difference between them. This effective input signal is obviously of lower amplitude than that applied at the base and so the gain of the amplifier is reduced due to the action of $R_3$. This is an example of what is called *negative feedback*.

The problem can be avoided, whilst still allowing $R_3$ to stabilise the operating point, by short-circuiting the emitter to the 0 V rail as far as a.c. signals are concerned. This is the function of $C_3$ in Fig. 5.13.

## 5.13 Establishing the operating point

The circuit of Fig. 5.13 was designed to give a quiescent collector current of 2 mA when using a silicon transistor with a typical $\beta$ of 100 and a nominal $V_{BE}$ of 0·6 V.

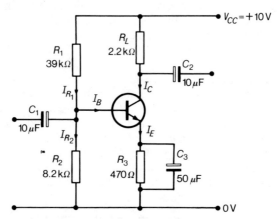

Fig 5.13 Completed amplifier. Resistor values calculated in text.

To achieve good thermal stability, the emitter voltage is commonly made 10% of the supply voltage. The supply here is 10 V and therefore $V_{R_3} = 1$ V

Since, $$I_E = 2 \text{ mA}$$ (Note $I_E \simeq I_C$)

$$R_3 = \frac{V_{R_3}}{I_E}$$

$$= \frac{1}{2 \times 10^{-3}} \Omega$$

$$= 500 \ \Omega$$ (Preferred value 470 Ω)

Now, $$I_B = \frac{I_C}{\beta}$$

*The Transistor Amplifier*

$$= \frac{2 \times 10^{-3}}{100} \text{ A}$$
$$= 20 \text{ μA}$$

So, making $I_{R_2}$ 10 times larger than the base current:

$$I_{R_2} = 10 I_B$$
$$= 10 \times 20 \times 10^{-6} \text{ A}$$
$$= 0\cdot2 \text{ mA}$$

Clearly,
$$V_{R_2} = V_{R_3} + V_{BE}$$
$$= 1 + 0\cdot6 \text{ V}$$
$$= 1\cdot6 \text{ V}$$

∴
$$R_2 = \frac{V_{R_2}}{I_{R_2}}$$
$$= \frac{1\cdot6}{0\cdot2 \times 10^{-3}} \text{ Ω}$$
$$= 8 \text{ kΩ} \qquad \text{(Preferred value } 8\cdot2 \text{ kΩ)}$$

The current in $R_1$ is the sum of $I_{R_2}$ and the base current.

∴
$$I_{R_1} = I_{R_2} + I_B$$
$$= 10 I_B + I_B$$
$$= 11 I_B$$
$$= 0\cdot22 \text{ mA}$$

The p.d. across $R_1$ is:

$$V_{R_1} = V_{CC} - V_{R_2}$$
$$= 10 - 1\cdot6 \text{ V}$$
$$= 8\cdot4 \text{ V}$$

∴
$$R_1 = \frac{V_{R_1}}{I_{R_1}}$$
$$= \frac{8\cdot4}{0\cdot22 \times 10^{-3}} \text{ Ω}$$
$$= 38\cdot2 \text{ kΩ} \qquad \text{(Preferred value } 39 \text{ kΩ)}$$

When choosing the quiescent collector voltage for the basic amplifier of Fig. 5.3(*a*), it was made equal to half the supply voltage. This allows equal positive and negative excursions of the output signal. A similar result is required here. However in Fig. 5.13, 1 V of the available 10 V supply is dropped across $R_3$ but this resistor is effectively short circuited by $C_3$ as far as a.c. signals are concerned. The effective collector to emitter supply for the transistor is now only 9 V and so, to permit equal maximum excursions of the output signal this time, $V_{CE}$ must be made 4·5 V. Since the emitter voltage is 1 V, the quiescent collector voltage is:

$$V_C = V_{R_3} + V_{CE}$$
$$= 1 + 4\cdot5 \text{ V}$$
$$= 5\cdot5 \text{ V}$$

The p.d. across $R_L$ is:

$$V_{RL} = V_{CC} - V_C$$
$$= 10 - 5\cdot5 \text{ V}$$
$$= 4\cdot5 \text{ V}$$

$\therefore$
$$R_L = \frac{V_{RL}}{I_C}$$
$$= \frac{4\cdot5}{2 \times 10^{-3}} \Omega$$
$$= 2\cdot25 \text{ k}\Omega \qquad \text{(Preferred value 2·2 k}\Omega\text{)}$$

The capacitors $C_1$ and $C_2$ are coupling capacitors which isolate the d.c. conditions of this stage from the adjacent stages. Like $C_3$ they are chosen to have negligible reactance at the signal frequencies. In the a.c. equivalent circuit for this amplifier, shown in Fig. 5.14, $C_1$ and $C_2$ are considered to be short circuited and the emitter joined directly to the common rail because of the presence of $C_3$. The effective resistance in parallel with the base and emitter (equivalent to $R_B$ for the basic amplifier) is now $R_1$ in parallel with $R_2$. The expressions for current gain, voltage gain and power gain developed earlier apply here also.

Hence the voltage gain for this amplifier, assuming $\beta = 100$, $r_{in} = 2 \text{ k}\Omega$ and $r_{out} \gg R_L$ is:

$$A_v = -\beta \frac{R_L}{r_{in}}$$
$$= -\frac{100 \times 2\cdot2 \times 10^3}{2 \times 10^3}$$
$$= -110$$

In order to calculate the current gain for this amplifier, the sharing of the input signal current between the base potential divider and the base must be taken into account. In the a.c. equivalent circuit, $R_1$ and $R_2$ must be considered to be in parallel (remembering the supply rails are common as far as a.c. signals are concerned).

# The Transistor Amplifier

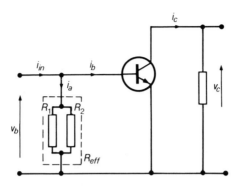

Fig 5.14 Simple a.c. equivalent circuit for the amplifier in Fig. 5.13.

Now,
$$R_{eff} = \frac{R_1 R_2}{R_1 + R_2}$$
$$= \frac{8 \cdot 2 \times 39}{8 \cdot 2 + 39} \text{ k}\Omega$$
$$\simeq 6 \cdot 8 \text{ k}\Omega$$

∴
$$i_b = \frac{R_{eff}}{r_{in} + R_{eff}} \times i_{in}$$
$$= \frac{6 \cdot 8}{2 + 6 \cdot 8} i_{in}$$
$$= 0 \cdot 77\, i_{in}$$

Therefore only 77% of the input signal current actually enters the base of the transistor. The current gain for the stage is therefore $0 \cdot 77\, \beta = 77$, and clearly the effect of the base resistors is not negligible in this case.

Finally, in this chapter the load resistance for the amplifiers has been the collector resistor $R_L$. In practice, if the amplifier output is connected to another circuit, e.g. the input of a second amplifier, the effective a.c. load resistance will be the collector resistor in parallel with the input resistance for the next stage. The effective a.c. load resistance, and hence the values of gain realised, will therefore be lower than those calculated here.

## Problems
(Answers at end of book)

1. An amplifier of the type shown in Fig. 5.1 is to be operated from a 9 V d.c. supply with a quiescent collector current of 1·6 mA. Determine the resistances $R_B$ and $R_L$ such that the maximum undistorted output signal voltage swing may be obtained.

Assume a silicon transistor with $\beta = 100$. Give the appropriate preferred values for the resistors.

2  Plot the common emitter static output characteristics for the transistor for which the data in the table below applies. The transistor is to be used in an amplifier of the type shown in Fig. 5.1. If the supply voltage is 12 V, and the load resistor $R_L$ is 3·3 kΩ, construct a load line on these characteristics.

If $R_B$ is chosen to give a base current of 20 μA, determine the quiescent operating point, $Q_1$.

If $R_B$ is changed to give a base current of 10 μA, determine the new operating point $Q_2$ and the maximum output signal that can be obtained at this operating point.

|  | $I_C$ (mA) |  |  |  |  |  |
| --- | --- | --- | --- | --- | --- | --- |
| $V_{CE}$ | 2 V | 4 V | 6 V | 8 V | 10 V | 12 V |
| $I_B = 10$ μA | 1 | 1·05 | 1·1 | 1·15 | 1·2 | 1·25 |
| $I_B = 20$ μA | 2 | 2·08 | 2·12 | 2·2 | 2·25 | 2·3 |
| $I_B = 30$ μA | 3 | 3·1 | 3·2 | 3·25 | 3·3 | 3·4 |

3  For a common emitter amplifier of the type shown in Fig. 5.13, determine the resistances $R_1$, $R_2$, $R_3$ and $R_L$ such that the maximum undistorted output signal voltage can be obtained. Assume a supply voltage of 10 V, a silicon transistor with $\beta = 100$ and a desired collector current of 1·5 mA. Give the corresponding preferred values for these resistors.

4  Calculate the approximate voltage gain for a common emitter amplifier with a collector load resistor of 2·7 kΩ if, at the quiescent operating point, $\beta = 120$ and $r_{in} = 1·8$ kΩ.

5  For the operating point $Q_1$ in question 2, deduce the short circuit current gain of the transistor. If the transistor has an input resistance of 2 kΩ, determine the approximate voltage gain of the amplifier.

Compare this result with that obtained by purely graphical methods using the load line plotted previously.

# Chapter 6
## L-C Oscillators

### 6.1

An $L$-$C$ oscillator is a sinusoidal waveform generator. The oscillator operation makes use of the properties of an $L$-$C$ circuit at resonance, and so we shall begin this chapter by examining the action of such a circuit.

### 6.2 Natural oscillations in an $L$-$C$ circuit

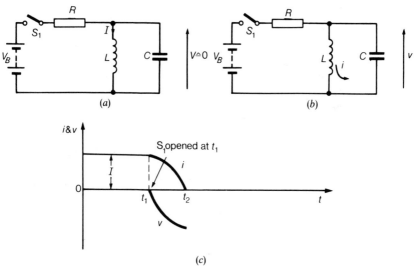

Fig 6.1 In the interval between $t_1$ and $t_2$ the collapsing magnetic flux in $L$ develops a current which charges $C$.

In Fig. 6.1(a), the switch $S_1$ is closed and so the battery is connected to the rest of the circuit. A direct current $I$ is therefore established. If the resistance of $L$ is negligible compared with $R$, the p.d. across $L$ (and of course $C$) may be assumed to be zero. The current in the inductor supports a magnetic field about it, the intensity of which is proportional to the magnitude of the current. This field stores an amount of energy equal to that initially used to develop the field when the switch was first closed.

When the switch is opened (shown as the instant $t_1$ in Fig. 6.1(c)) the current in $L$ starts to decrease. The magnetic field supported by the current also decreases and, as it does so, it cuts the turns of $L$ inducing an e.m.f. into them. The polarity of this induced e.m.f. is such that it attempts to maintain a current in the same direction as before. The necessary energy is drawn from the magnetic field and the current decreases progressively as the field collapses. The circuit for this current is completed by the capacitor $C$. The capacitor therefore acquires charge and a p.d. is developed between its plates. Fig 6.1(c) shows, during the interval $t_1$ to $t_2$, the decreasing current and the build up of p.d. across the capacitor. Note that the direction of the current is such that the upper plate of the capacitor becomes negative with respect to the lower plate.

At the instant $t_2$, the field in $L$ has completely collapsed and the current is zero. The situation is now one of a charged capacitor with $L$ completing a circuit between its plates. The capacitor therefore discharges via the inductor, developing a current in the opposite direction to the original current. The way in which the current and p.d. change whilst the capacitor is discharging is shown in the interval $t_2$ to $t_3$ in Fig. 6.2. Note that as the capacitor discharges, the rate of growth of current gradually decreases.

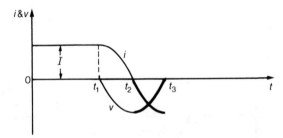

Fig 6.2 In the interval between $t_2$ and $t_3$ $C$ discharges back through $L$.

At the instant $t_3$, the capacitor is completely discharged and the p.d. between its plates is zero. The current, at this time, has its maximum negative value and the energy is again stored in the magnetic field about the inductor. Since the p.d. across the capacitor is zero, the current now starts to decrease but, as before, the e.m.f. induced in the inductor by the collapsing magnetic field attempts to maintain this current. The capacitor forms a circuit for the current and so once again it acquires charge. This time however, because the direction of the current has reversed compared with that during the interval $t_1$ to $t_2$, this charge has the opposite polarity to that which it held previously. The

# L-C Oscillators

changes in current and p.d. whilst the capacitor is being recharged in this new direction are shown in the interval $t_3$ to $t_4$, in Fig. 6.3. Note that as the current produced by $L$ decreases, the rate of growth of p.d. across the capacitor decreases also.

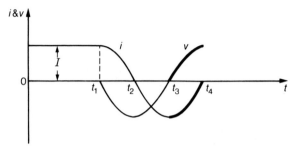

Fig 6.3 In the interval between $t_3$ and $t_4$, the collapsing flux in $L$ develops a current which charges $C$.

From the events described so far, it is evident that the energy originally stored by the magnetic field of the inductor is being repeatedly exchanged between the inductor and capacitor. When the inductor is charging the capacitor, the energy is being transferred from magnetic storage to electrostatic storage, and vice versa. The electric current in the circuit is the means by which the energy exchanges take place. This current, which has a continually changing polarity and rate of growth, is found to be sinusoidal in form. The p.d. developed across the circuit by this current is also, of course, sinusoidal. The magnitude of this p.d. gradually decreases because, during each exchange, energy is dissipated in the unavoidable resistive losses of the circuit. A more complete sketch of these decaying *oscillations* (or *damped* oscillations as they are known) is shown in Fig. 6.4.

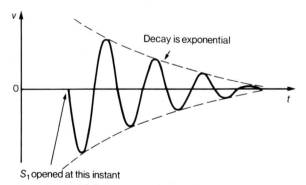

Fig 6.4 Damped oscillations in the $LC$ circuit.

The natural oscillations occur at a particular frequency (for any given values of L and C) and the circuit is referred to as a *resonant* or *tuned* circuit. The frequency of the oscillations $f_0$, can be calculated from the expression:

$$f_0 = \frac{1}{2\pi\sqrt{LC}} \text{ Hz}$$

## 6.3 Sustained oscillations

If energy can be supplied to the L-C circuit to make up for that dissipated in the resistive losses, the amplitude of the oscillations can be maintained at a constant level. This energy can be injected into the circuit say, once per cycle. The amount injected must, or course, be equal to the total loss of energy over the whole of each cycle. The timing of this operation must be such that the additional e.m.f. introduced into the circuit enhances the natural oscillation by the desired amount. If this is achieved, the sustained oscillations shown in

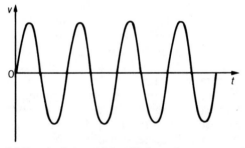

Fig 6.5 Sustained sinusoidal oscillations of constant amplitude.

Fig. 6.5 will result. A circuit which is designed to maintain the oscillations in a tuned circuit, and thereby provide a sinusoidal output of stable amplitude, is considered in the next section.

## 6.4 Tuned collector oscillator

Figure 6.6 shows the circuit of a *tuned collector* oscillator. This circuit will provide an output of nominally constant frequency and amplitude. A tuned circuit formed by $L_1$ and $C_1$, is connected as the collector load of $TR_1$. A second inductor $L_2$, is mutually coupled to $L_1$. As far as a.c. signals are concerned, $L_2$ is connected between the base and emitter of $TR_1$ via the low reactance of the capacitors $C_2$ and $C_3$.

When the supply is first connected, a transient current is developed in the tuned circuit as the collector current rises to its initial quiescent value. This transient current initiates natural oscillations in the tuned circuit. These natural oscillations induce a small e.m.f. into $L_2$ and hence cause corresponding variations in the base current. The phase of the e.m.f. induced into $L_2$, is such that when the oscillations in the tuned circuit cause the collector voltage to fall, the base current of the

# L-C Oscillators

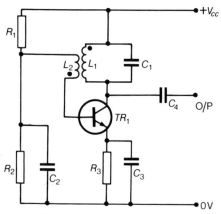

Fig 6.6 Tuned collector oscillator.

transistor is simultaneously increased. The increase in the base current causes an increase in the collector current and, because of the additional p.d. that this develops across the impedance of the tuned circuit, further lowers the collector voltage. Since the increase in collector current assists the change in p.d. which the natural oscillations were producing, it is evident that the increase in collector current results in energy being added to the tuned circuit.*

When the oscillations in the tuned circuit cause the collector voltage to rise, the phase of the e.m.f. which is now induced into $L_2$ is such that the base current is cut off (this situation is more thoroughly dealt with in section 6.5). This allows the collector voltage to rise without current being drawn from the tuned circuit which would otherwise damp this half-cycle of the natural oscillations.

The correct phase relationship of the signal at the base of the tran-

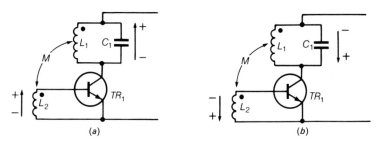

Fig 6.7 (a) Showing the phase of the e.m.f. induced in $L_2$ when the collector voltage falls.
(b) Showing the phase of the e.m.f. induced in $L_2$ when the collector voltage rises.

---

* The process described in this paragraph is an example of positive feedback. The general principles of positive feedback are described in section 6.6.

sistor, with respect to the signal at its collector, is established by the relative connections of the windings $L_1$ and $L_2$. The phase of the e.m.f. induced in $L_2$ when the natural oscillations in the tuned circuit result in a decrease in the collector voltage, is illustrated in Fig. 6.7(a). For the alternate half-cycle both signals are reversed as in Fig. 6.7(b).

The frequency of oscillation will be approximately equal to the resonant frequency of the tuned circuit. That is:

$$f_0 \simeq \frac{1}{2\pi\sqrt{L_1 C_1}} \text{ Hz}$$

### 6.5 Amplitude self-stabilisation—class C operation

For a short time after the connection of the supply, the resistors $R_1$, $R_2$ and $R_3$ bias the transistor into class $A$. In other words, there is a collector current of sufficient magnitude for the transistor to operate as a linear amplifier. The sinusoidal oscillation promoted by the initial rise in collector current can therefore, by means of the e.m.f. induced in $L_2$, cause a sinusoidal component of collector current. This action should lead to the maintenance of oscillations.

In order to ensure that continuous oscillations will occur under a variety of conditions, it is necessary to design the oscillator circuit such that under starting conditions, more energy is injected into the tuned circuit per cycle than is lost per cycle. That is, the amplifier gain is made greater than that necessary to sustain oscillations. This is necessary because the amplification provided by the transistor may vary with both temperature and supply voltage. For example, if the circuit was designed such that it had just sufficient gain to sustain oscillations at room temperature, then under colder conditions it may not oscillate at all. Injecting more energy into the tuned circuit than is required to make up for the losses will, if unchecked, lead to a progressive increase in the amplitude of oscillation. However, when the oscillation grows to a sufficient magnitude, the signal induced in $L_2$ will begin to cut off the transistor. That is, during the negative half-cycle at the base of the transistor, the base current is cut off. This produces a situation which, in effect, represents rectification of the signal at the base, the rectifying action being provided by the base emitter junction.

Figure 6.8 illustrates the situation during the positive half-cycle at the base, that is, when the base-emitter junction conducts. The resultant current $i_b$ adds charge to $C_2$. The direction of this base current being such that the base voltage is made less positive with respect to the 0 V line. At the same time a much larger current $i_e$ (larger due to the current amplification of the transistor) adds charge to $C_3$. This causes the emitter voltage to become more positive with respect to the 0 V line. Now only a small amount of the additional charge acquired by these two capacitors, will be removed during the negative half-cycle

# L-C Oscillators

when the transistor is switched off ($C_3$ discharges via $R_3$, and $C_2$ loses its additional charge by virtue of $R_1$ and $R_2$). This means that there is a net change in the direct voltages at the base and the emitter. Both changes (that is an increase of emitter voltage and a decrease of

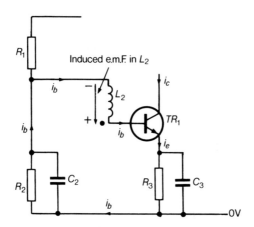

Fig 6.8 Illustrating the situation when the e.m.f. induced in $L_2$ causes the base-emitter junction to conduct.

base voltage) tend to cause the transistor to cut off. In fact, these voltages can change sufficiently for the transistor to be cut off for a large part of each cycle. The base-emitter junction is then forward biased by only the tip of the positive half-cycle of the signal at the base, resulting in just a pulse of collector current once per cycle. This situation is illustrated by the waveforms of Fig. 6.9.

The bias will automatically adjust itself such that this pulse of collector current injects just sufficient energy to compensate the losses in the tuned circuit. The amplitude of oscillations will therefore remain virtually constant. Should the amplitude of oscillations increase however, the amplitude of the signal in $L_2$ and hence the magnitude of the base-emitter reverse bias will also increase. This tends to cut the transistor off for an even larger part of each cycle. The reduced collector current pulse is not then sufficient to maintain this increased level of oscillation and it falls back to the previous amplitude. Similarly, if the amplitude of oscillation tends to decrease, the collector current pulse is increased and the amplitude of oscillation rises again. This is the inherent amplitude control action of the circuit which ensures that the amplitude of oscillation remains reasonably constant.

The mode of operation, where the transistor conducts for much less than 180° of any complete cycle at its input, as here, is known as *class C* operation.

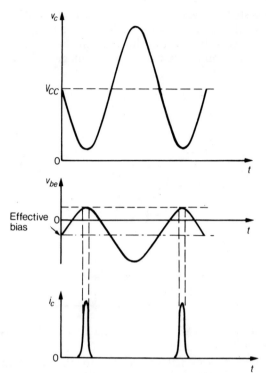

Fig 6.9 Signal waveforms in the tuned collector oscillator after the build up of oscillations.

## 6.6 Positive feedback

The tuned collector oscillator is essentially an amplifier which has part of its output signal fed back to its input circuit. The phasing of this signal is such that the amplifier provides its own input signal, which in turn, sustains the output. This is an example of *positive feedback*.

Consider the system diagram of Fig. 6.10 in which an amplifier of gain $A$, has a circuit connected to its output such that a fraction $\beta$ of the output signal is delivered to point $y$.

If the signal developed at point $y$ is identical in phase to the input signal, then connecting point $y$ to the input would clearly enlarge the input signal and result in a greater output. This situation would constitute positive feedback. It can be seen therefore, that for positive feedback, there must be no net phase shift in the closed loop formed by the amplifier and the feedback circuit.

Referring once again to Fig. 6.10, assume that the signal at point $y$ is not only identical in phase to the input signal, but also equal in amplitude to the input signal. It would under these circumstances, be possible to connect point $y$ to the input and remove the applied test

# L-C Oscillators

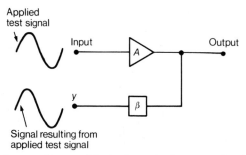

Fig 6.10 System diagram of an amplifier having a circuit connected to its output such that a fraction $\beta$ of the output signal is delivered to point 'y',

signal with no resultant change in the output. The circuit can then provide a continuous sinusoidal output without having any external signal input. It will in fact be a sinusoidal oscillator. For continuous oscillations therefore, the total gain around the closed loop formed by the amplifier and the feedback circuit must be equal to (or greater than) unity. In Fig. 6.10, when point y is joined to the input, the gain around the feedback loop (the *loop gain*) is equal to the product of $A$ and $\beta$. Therefore, for continuous oscillations $\beta A \geq 1$.

The two conditions necessary for continuous oscillation, *zero loop phase shift* and *unity loop gain*, are met by the circuit in Fig. 6.6 in the following manner. Zero loop phase shift is provided by $TR_1$ introducing 180° phase shift, followed by a feedback circuit (feedback from collector to base via the mutual coupling between $L_1$ and $L_2$) which introduces a further 180° phase shift due to the way that $L_2$ is connected into the base circuit. That is, when the collector end of $L_1$ is going positive, the end of $L_2$ which is connected to the base is going negative and vice versa. Two phase shifts of 180° in the loop, result in an overall phase shift of zero. In order for the common emitter amplifier to provide a 180° phase shift it needs a resistive collector load. At the frequency at which the natural oscillations occur in the $LC$ circuit (the resonant frequency), it acts like a pure resistance and therefore meets the requirement. The unity loop gain requirement is met by the amplification which the transistor provides, compensating the attenuation which the feedback circuit introduces. In order to ensure the commencement of oscillations, the circuit is designed to have a loop gain greater than unity when working in its initial class $A$ mode. This builds up the amplitude of oscillations until they are finally limited by the circuit transferring to the class $C$ mode described in section 6.5.

## 6.7 The Hartley oscillator
Figure 6.11 shows the circuit of a Hartley type oscillator. The general principle of operation of this circuit is similar to that of the tuned collector oscillator examined previously. Oscillations in the tuned cir-

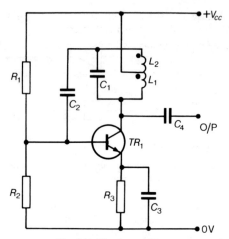

Fig 6.11 Hartley oscillator.

cuit, which forms the collector load of $TR_1$, are maintained by the transistor controlling the supply of energy to the circuit as before.

A portion of the oscillatory p.d. across the L-C circuit, (that developed in $L_2$), is fed back to the base-emitter circuit of $TR_1$. Note (since the d.c. supply will have negligible impedance to a.c.), for signal purposes the lower end of $L_2$ (i.e. the tap on the inductor) can be considered to be connected to the emitter. The top end of $L_2$ is connected to the base via the negligible reactance of $C_2$.

$TR_1$ is connected in the common emitter mode and will therefore introduce a 180° phase shift inside the feedback loop. The second 180° phase shift which is required to achieve zero loop phase shift, is provided by the way in which $L_2$ is connected to the base-emitter circuit. The circuits in Fig. 6.12(a) and (b) represent the effective situation as

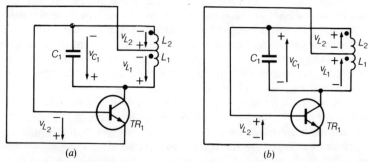

Fig 6.12 (a) Effective signal circuit of the Hartley oscillator showing the phase of the feedback signal between the base and emitter, during the positive half-cycle at the collector.

(b) Phase of the signal between the base and emitter during the negative half-cycle at the collector.

# L-C Oscillators

far as a.c. signals are concerned. The distribution of p.d. across the L-C circuit is shown for each of the oscillatory half-cycles. Note that the signal at the base which results from the signal at the collector is phase inverted.

The frequency of oscillation will be approximately equal to the resonant frequency of the L-C circuit. The total inductance $L'$ of the tuned circuit is:

$$L' = L_1 + L_2 + 2M$$

where $M$ is the mutual inductance between $L_1$ and $L_2$

Therefore the frequency of oscillations:

$$f_0 \simeq \frac{1}{2\pi\sqrt{L'C}} \text{ Hz}$$

## 6.8 The Colpitts oscillator

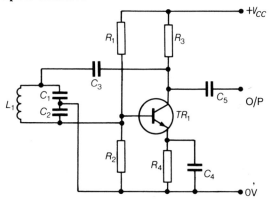

Fig 6.13 Colpitts oscillator.

Figure 6.13 shows a Colpitts type oscillator circuit. Once again the general principle of operation of the circuit is the same as that of the tuned collector oscillator. Oscillations in the L-C circuit formed by $L_1$, $C_1$ and $C_2$, are maintained by $TR_1$ controlling the energy input to the circuit.

The essential difference between this circuit and the Hartley oscillator of section 6.7, is the way in which the feedback signal is derived from the L-C circuit. The portion of the oscillatory p.d. that is developed across $C_2$ is applied to the base-emitter circuit of $TR_1$ (in effect this circuit uses a capacitive tapping whereas the Hartley oscillator uses an inductive tapping).

The 180° phase shift introduced into the feedback loop by the transistor is complemented by the 180° phase shift obtained by the way in which the feedback signal is connected to the base-emitter circuit. The circuits in Fig. 6.14 represent the effective situation as far as the a.c. signals are concerned. The distribution of p.d. across the $L$-$C$ circuit is shown for each half-cycle. Note that the signal at the base which results from the signal at the collector is phase inverted. The function of $C_3$ is to prevent an effective d.c. short circuit between the collector and base of $TR_1$ via the inductor $L_1$. This capacitor will be chosen to have a negligible reactance to the oscillatory frequency and so is not included in the diagrams of Fig. 6.14.

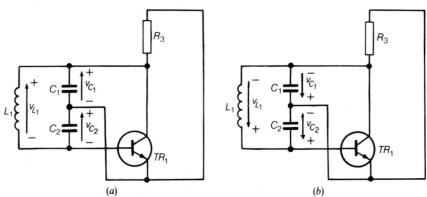

Fig 6.14 (a) Effective signal circuit of the Colpitts oscillator showing the phase of the feedback signal between the base and emitter, during the positive half-cycle at the collector.
(b) Phase of the signal between the base and emitter during the negative half-cycle at the collector.

The frequency of oscillation will be approximately equal to the resonant frequency of $L$-$C$ circuit. The total capacitance $C'$ of the $L$-$C$ circuit is formed by $C_1$ and $C_2$ in series:

$$C' = \frac{C_1 C_2}{C_1 + C_2}$$

The frequency of oscillation is then:

$$f_0 \simeq \frac{1}{2\pi\sqrt{L_1 C'}} \text{ Hz}$$

**Problems**
(Answers at end of book)

1  (a) With the aid of a simple system diagram, describe the two conditions that are necessary for the maintenance of continuous oscillations.
(b) Describe how these two conditions are met in (i) a Colpitts oscillator and (ii) a Hartley oscillator.

# L-C Oscillators

2 (a) Draw the circuit of a tuned collector oscillator and with reference to this circuit explain how, after the supply is first connected, it transfers from class $A$ to class $C$ operation.

(b) Explain how the circuit action which brings about class $C$ operation results in amplitude control of the oscillations.

3 (a) A Hartley oscillator of the type shown in Fig. 6.11 has a tuned circuit having the following component values: $C_1 = 50$ nF, $L_1 = 200$ $\mu$H, $L_2 = 2$ $\mu$H. The mutual inductance between $L_1$ and $L_2$ is 10 $\mu$H. Determine the output frequency of the oscillator.

(b) Determine the approximate frequency of operation ignoring $L_2$ and the mutual inductance.

4 (a) A Colpitts oscillator of the type shown in Fig. 6.13 has the following component values for its tuned circuit: $L_1 = 80$ $\mu$H, $C_1 = 50$ nF, $C_2 = 500$ nF. Determine the frequency of oscillation.

(b) Determine the approximate frequency of oscillation ignoring $C_2$.

# Chapter 7
# Waveform Generators

*L-C* oscillators, which are of course waveform generators which produce a sine wave signal, were dealt with in the previous chapter. In this chapter, we shall consider some waveform shaping circuits and other circuits which may be used to generate square, rectangular and sawtooth output signals. Firstly, we shall examine the charge and discharge action of a capacitor-resistor network as this is an important aspect of the operation of these circuits.

## 7.1 *C-R* network
Consider the *C-R* circuit in Fig. 7.1(*a*). The switch is open and consequently there is no current. The p.d. across the resistor is therefore zero. Assuming that the capacitor is initially uncharged, the p.d. across it is zero also.

If the switch is now moved to position 2 (Fig. 7.1(*b*)), the capacitor will begin to charge. It cannot charge instantaneously however, because the charging current is limited by the resistor. At the instant the switch is closed there is no p.d. across the capacitor, and so the full supply p.d. of 10 V appears across the resistor. The initial charging current is therefore:

$$i_1 = \frac{v_R}{R}$$
$$= \frac{V_S}{R} \quad \text{(since, when } t = 0, v_R = V_S\text{)}$$
$$= \frac{10}{10 \times 10^3} \text{ A}$$
$$= 1 \text{ mA}$$

A short time later, the capacitor will have acquired some charge and the p.d. across it will have risen to say 2 V. The p.d. across the resistor is then:

$$v_R = V_S - v_C$$

$$= (10 - 2) \text{ V}$$
$$= 8 \text{ V}$$

The charging current at this instant is therefore:

$$i_1 = \frac{v_R}{R}$$
$$= \frac{8}{10 \times 10^3} \text{ A}$$
$$= 0{\cdot}8 \text{ mA}$$

Note, the charging current has decreased and so the capacitor is now charging more slowly. This process is progressive, the rate of charge decreasing as the p.d. across the capacitor approaches the applied voltage.

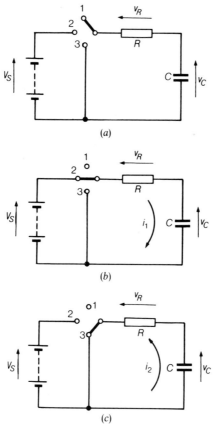

Fig 7.1 Charge and discharge of a capacitor in a $C$–$R$ network. $V_S = 10$ V, $R = 10$ k$\Omega$, $C = 0{\cdot}1$ $\mu$F.

The p.d. across the capacitor rises exponentially with time as illustrated in Fig. 7.2(*a*). The p.d. across the resistor, which is at all times equal to the difference between the applied voltage and the p.d. across the capacitor, must therefore decrease in the manner shown in Fig. 7.2(*b*). The decreasing charge current is illustrated in Fig. 7.2(*c*).

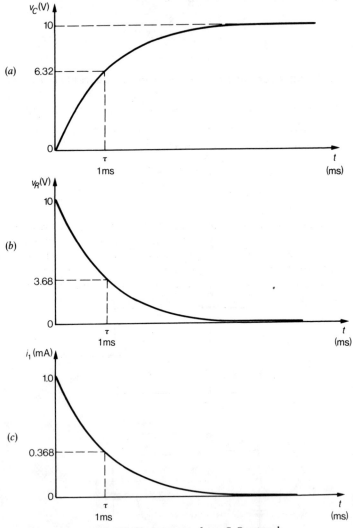

Fig 7.2 Charge curves for a *C–R* network.

It may be shown, for a *C-R* network, that the growth of the p.d. across the capacitor is given by:

$$v_C = V_S(1 - e^{\frac{-t}{CR}}) \qquad (1)$$

where $t$ is the time elapsed since the closing of the switch.

The product $CR$ in the exponential term is called the time constant $\tau$ of the circuit (i.e. $\tau = CR$).

Using the values in Fig. 7.1 the time constant is:

$$\begin{aligned}\tau &= CR \\ &= 0\cdot 1 \times 10^{-6} \times 10 \times 10^3 \text{ s} \\ &= 1 \text{ ms}\end{aligned}$$

Substituting this result, and also $V_S = 10$ V, in equation (1) we get:

$$v_C = 10(1 - e^{\frac{-t}{10^{-3}}}) \text{ V}$$

Taking the particular case where $t$ (the time elapsed since the closing of the switch) is equal to the time constant (i.e. $t = 1$ ms), then:

$$\begin{aligned}V_c &= 10(1 - e^{-1}) \text{ V} \\ &= 10(1 - 0\cdot 368) \text{ V} \\ &= 6\cdot 32 \text{ V}\end{aligned}$$

Therefore, the p.d. across the capacitor rises to 63·2% of the applied voltage after a time equal to the time constant of the C-R network.

The complete exponential curve representing the growth of p.d. across the capacitor can be plotted directly by inserting various values of $t$ in equation (1). However, using the above result, the curve can be sketched more simply as in Fig. 7.3.

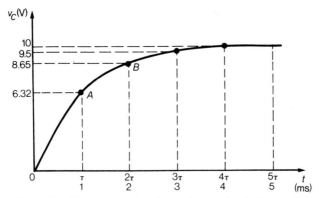

Fig 7.3 Simplified construction of growth curve for C–R network.

The supply is 10 V, and the capacitor charges to 63·2% of this (6·32 V) in a period equal to one time constant (1 ms); marked point A. This can be considered to be a new starting point. The capacitor will now charge additionally by 63·2% of the remaining p.d. during the next time constant period. After a period equal to two time constants the p.d. across the capacitor is:

$$v_C = 6{\cdot}32 + \left[\frac{63{\cdot}2}{100} \times 3{\cdot}68\right] \text{V}$$
$$= 8{\cdot}65 \text{ V (Point } B)$$

During the period $2\tau$ to $3\tau$ the capacitor will further charge by 63·2% of the remaining p.d. and so on. For practical purposes, we may assume the capacitor to be fully charged after a period equal to $3\tau$ (95% of applied voltage) or, if greater precision is required, after $5\tau$ (99·3% of applied voltage).

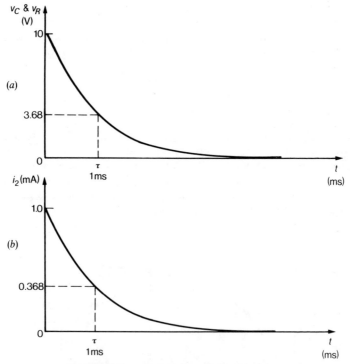

Fig 7.4 Discharge curves for a $C$–$R$ network.

Theoretically, as the technique just used indicates, the capacitor never charges completely since the remainder just gets smaller and smaller. In practice however, the capacitor will pass a small leakage current. For a given applied voltage therefore, the minimum current cannot become any smaller than this leakage current.

Now assume the capacitor has fully charged and the switch is moved to position 3, as illustrated in Fig. 7.1(c). The capacitor will now discharge through the resistor. Note that, in this position, the supply is disconnected and the resistor is placed in parallel with the capacitor. Initially, the p.d. across $R$ and $C$ is 10 V. The initial discharge current is therefore:

$$i_2 = \frac{v_{c(\max)}}{R}$$

$$= \frac{10}{10 \times 10^3} \text{ A}$$

$$= 1 \text{ mA}$$

The p.d. across the capacitor, and hence the p.d. across the resistor, will decrease as the capacitor discharges. The nature of the discharge is an exponential decay as shown in Fig. 7.4(a). The capacitor discharges by 63.2% of the initial p.d. during a time equal to one time constant. It discharges by 63.2% of the remainder in the next time constant period, and so on. Since the resistor and capacitor are in parallel, the p.d. across the resistor decreases in the same way. The discharge current $i_2$ is proportional to $v_R$, and so it decreases in an equivalent manner, as in Fig. 7.4(b).

## 7.2 C-R circuits

Now consider how a C-R network behaves when a square wave signal is applied. This is illustrated in Fig. 7.5. The input is a unidirectional square wave of 10 V amplitude. When the input voltage steps up to $+10$ V (a situation equivalent to closing the switch in Fig. 7.1(b)), the capacitor starts to charge. The time constant of the C-R network is again 1 ms. The duration of the pulse is 5 ms. That is, the duration of the pulse is equivalent to 5 time constants (i.e. $\tau = \frac{t}{5}$). The capacitor can therefore be assumed to charge completely during the period the pulse is present.

After 5 ms the input voltage falls to zero. Taking the input to zero is equivalent to short circuiting the input terminals (as in Fig. 7.1(c)).

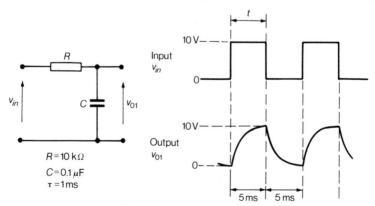

Fig. 7.5 Integrator with $\tau = t/5$.

Since the interval between pulses is 5 ms, the capacitor can be assumed to discharge completely during this period (5 time constants). The resultant output waveform has a sawtooth like appearance, as shown. This circuit arrangement is commonly called an integrating network or *integrator*.

An alternative circuit arrangement is shown in Fig. 7.6. The output this time is the p.d. across the resistor. Again starting with the capacitor uncharged, consider what happens at the instant the input voltage steps up to $+10$ V. Since the capacitor cannot charge instantaneously, and hence the p.d. across it cannot rise suddenly, the full 10 V is developed across the resistor and the charging current is a maximum. As the capacitor charges, the charging current decreases exponentially, and the output voltage decreases in the same manner (the output voltage being proportional to the charging current). After 5 ms, that is after a period equivalent to 5 time constants, the capacitor may be assumed to be completely charged. The charging current is then zero and likewise the output voltage is zero also.

The input voltage now steps down from $+10$ V to zero. Since the capacitor cannot discharge instantaneously, the output voltage must also step down by 10 V. That is, the output voltage steps to $-10$ V. As the capacitor discharges, the discharge current diminishes exponentially. The output voltage therefore changes in a similar fashion as it returns to zero.

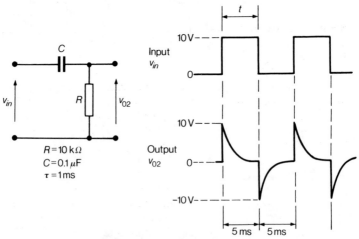

Fig 7.6 Differentiator with $\tau = t/5$.

It is evident therefore, that the output voltage has a peak to peak value which is twice the amplitude of the input square wave. Notice also that the ouput signal has a positive spike for each rising edge of the input signal, and a negative spike for each falling edge of the input

# Waveform Generators

signal. This circuit arrangement is commonly called a differentiating network or *differentiator*.

Although the output waveforms for the integrator and differentiator are quite different it should be recognised that the two circuits, when viewed at the input terminals, are essentially the same, i.e. a resistor and capacitor in series. The component values, and the input waveform are also the same in the two cases considered here. The voltage waveform across the resistor in each circuit will therefore be identical. Likewise, the voltage waveform across the capacitors will be the same in each case. Now, remember that the sum of the potential differences across the resistor and the capacitor must equal the applied p.d. at all times. The sum of $v_{O_1}$ (in Fig. 7.5) and $v_{O_2}$ (in Fig. 7.6) must therefore be a square wave signal equivalent to the input signal. It is instructive to recognise this relationship as the nature of output waveforms to be expected from related circuits becomes more apparent, as seen in the next example.

In Fig. 7.7 an integrator with a smaller time constant is considered, i.e. $\tau = \frac{t}{10}$. The capacitor can be considered to be fully charged after a period of $5\tau$, and then the output will remain constant until the input changes again. The output waveform is therefore very much more square in shape.

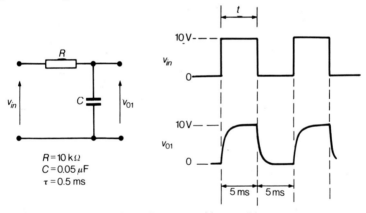

Fig 7.7 Integrator with $\tau = t/10$.

The output from the corresponding differentiator (i.e. $\tau = \frac{t}{10}$ again) in Fig. 7.8 can be deduced by considering the current in the circuit as before. Equally, it may be deduced by inspection of Fig. 7.7. That is, the time constants of the integrator and differentiator circuits and their input signals are the same. Consequently, the output from the differentiator is again equivalent to the difference between the input and output waveforms of the integrator. Note again, that there is a positive spike for each rising edge and a negative spike for each falling edge of the input waveform.

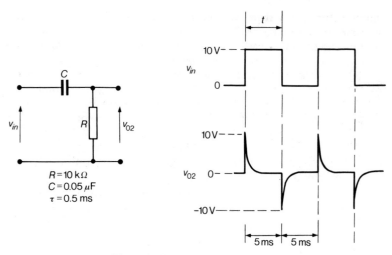

Fig 7.8 Differentiator with $\tau = t/10$.

## 7.3 Simple ramp generator

A *C-R* network may be used to produce a voltage ramp as shown in Fig. 7.9. This could form the basis for a simple timebase for a cathode ray oscilloscope or a television receiver tube, although the linearity of the ramp obtained by this method is seldom good enough for the former.

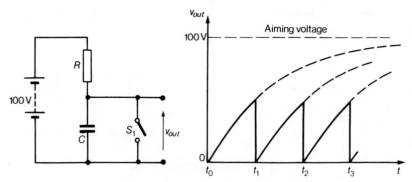

Fig 7.9 Simple voltage ramp generator.

Assume that the switch is first opened at the time $t = t_0$. The output voltage $v_{out}$ starts at zero and rises as the capacitor charges towards the aiming voltage (i.e. the supply of 100 V). At time $t_1$, the switch is momentarily closed and the capacitor is discharged. The switch is then opened again and the capacitor begins to charge once more. If the switch is closed at regular intervals (at $t_2$, $t_3$, etc.), the output waveform will have the sawtooth appearance shown. Note that, since only a small section of the capacitor charge curve is used, the ramp is reasonably linear. To make a practical timebase, the switch $S_1$ would have to be an electronic switch, e.g. a transistor as shown later in Fig. 7.19.

## 7.4 Clipper
It is sometimes desirable to generate short duration pulses corresponding to either the positive going (rising) or negative going (falling) edges of a square wave signal. These pulses may be required to trigger a monostable multivibrator for example (see section 7.5). Figure 7.10 shows one way in which such pulses may be produced. With this particular circuit, output pulses will be obtained for only positive going

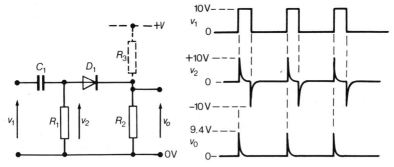

Fig 7.10 Diode clipper. Waveforms shown apply when $R_3$ is not connected.

edges of the input signal. The circuit uses a differentiator $C_1 R_1$ and a diode as a clipper. With no input signal applied the diode is unbiased, the anode and cathode being returned to the 0 V line via $R_1$ and $R_2$ respectively (ignore the presence of $R_3$ for the time being). The input signal is differentiated by $C_1 R_1$ to produce positive pulses corresponding to its positive going edges and negative pulses corresponding to its negative going edges. During the negative pulses the diode is reversed biased and so these pulses are blocked. When the positive pulses exceed say 0·6 V (p.d. sufficient to turn on the diode), the diode will conduct and these pulses will be transmitted to the output. The circuit therefore provides pulses which correspond to the positive going edges of the input signal only.

A positive bias can be applied to the diode cathode by introducing $R_3$. Let us assume that the resistances $R_2$ and $R_3$ are chosen to make the cathode voltage say +3 V. The positive spikes at the anode will now have to exceed about 3·6 V to turn the diode on. Only the tops of these positive pulses will now be transmitted to the output. This latter arrangement has the advantage that small noise signals carried on the input signal will not produce an output.

## 7.5 The monostable multivibrator
The monostable multivibrator (one-shot) circuit produces a rect-

angular output pulse in response to a trigger signal applied at its input. The duration of the output pulse may be set as required and this makes the circuit useful as a timer (e.g. to set the time delay between two operations). The relationship between the input and output waveforms for the circuit in Fig. 7.11(a) is illustrated in Fig. 7.11(b). Notice that an output pulse is produced each time a trigger pulse is applied. The duration of the output pulse $t_p$ is the same each time.

The circuit has two distinct operating states. One of these states is a stable condition (hence mono-stable) in which $TR_1$ is OFF and $TR_2$ is ON. This is the condition between output pulses. The other state is a quasi-stable state in which $TR_1$ is ON and $TR_2$ is OFF. This lasts for the duration of the output pulse $t_p$.

Consider first the stable state. This is the condition in which the

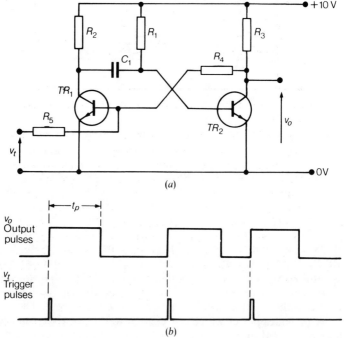

Fig 7.11 (a) Monostable multivibrator.
       (b) Waveforms showing the relationship between trigger pulses and output pulses.

circuit rests until a trigger pulse is applied. Notice that $TR_2$ base is connected to the positive supply line via $R_1$, whereas the base of $TR_1$ is connected to the collector of $TR_2$ via $R_4$. The resistance of $R_1$ is chosen so that the base current it provides is sufficient to saturate $TR_2$ (i.e. turn it hard on). The collector voltage of $TR_2$ is consequently very low, say 0·1 V. Since this 0·1 V, via $R_4$, represents the supply for $TR_1$ base,

# Waveform Generators

circuit rests until a trigger pulse is applied. Notice that $TR_2$ base is $TR_1$ will not conduct (i.e. it is turned off).* This is a stable condition which will persist until a trigger pulse is applied.

In this stable condition, the base voltage of the conducting transistor $TR_2$ will be say $+0.6$ V. The collector voltage of the non-conducting transistor $TR_1$ will be $+10$ V (since there is no current in $R_2$ and hence no p.d. across it). The p.d. across the capacitor $C_1$ is therefore 9·4 V with the polarity shown in Fig. 7.12.

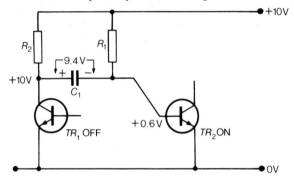

Fig 7.12 Section of monostable multivibrator circuit showing the p.d. across $C_1$ in the stable condition.

Now consider what happens when a trigger pulse is applied. In this circuit, the trigger pulses are positive pulses applied at the base of $TR_1$ (the non-conducting transistor).† When a trigger pulse is applied, it turns on $TR_1$ and drives it rapidly into saturation. The collector voltage of $TR_1$ quickly falls to 0·1 V. The capacitor $C_1$ cannot discharge instantaneously and so the base of $TR_2$ is driven negative. This is illustrated in Fig. 7.13 which shows the conditions at the instant $TR_1$ turns on. The capacitor has not yet started to discharge and there is still 9·4 V across it, the polarity being the same as before. However, there is only $+0.1$ V at $TR_1$ collector, and so the base of $TR_2$ must be at $-9.3$ V. $TR_2$ is therefore switched off.‡ This is the start of the quasi-stable condition with $TR_1$ on and $TR_2$ off. With $TR_2$ turned off, $TR_1$ is

---

* In practice $TR_1$ base may additionally be connected, via a resistor, to a negative supply. The base of $TR_1$ in this condition will then be negative with respect to the emitter which ensures that $TR_1$ is switched off.

† The trigger pulses may alternatively be negative pulses applied at the base of $TR_2$ (the conducting transistor). The trigger pulse will then switch off $TR_2$, and hence turn on $TR_1$, promoting the same action.

‡ A regenerative (positive feedback) action is involved in this switching process. When the trigger pulse is applied, it starts to turn on $TR_1$ and its collector voltage falls. $C_1$ cannot discharge instantaneously and so this decrease is transmitted directly to the base of $TR_2$. This reduces the collector current of $TR_2$ causing its collector voltage to rise. This in turn will further increase the base current of $TR_1$. The net result is that, very rapidly, $TR_1$ switches on and $TR_2$ switches off as described. A similar regenerative action will apply when the circuit switches back to the stable state.

provided with base current from the supply via $R_3$ and $R_4$. $TR_1$ will therefore continue to conduct after the trigger pulse has ended. In fact, this condition will persist until $C_1$ is able to discharge allowing the base of $TR_2$ to go positive again. This will allow $TR_2$ to conduct once more. $TR_1$ will then be deprived of base current, causing it to turn off, and the circuit will return to the stable state again.

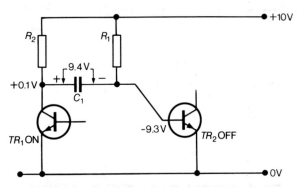

Fig 7.13 Section of monostable multivibrator showing the situation at the instant the circuit switches to the quasi-stable state.

Now consider how the period of the quasi-stable state, and hence the duration of the output pulse, is determined. The output pulse is obtained from $TR_2$ collector and represents the period for which $TR_2$ is off. This in turn is determined by how long it takes for $C_1$ to discharge. The discharge path for $C_1$ is via $R_1$ and $TR_1$. $TR_1$ can be regarded simply as a switch which connects the left-hand plate of $C_1$ to the 0 V rail (the p.d. across $TR_1$ is only say 0·1 V when it is switched on). The duration of the output pulse is therefore governed by the time constant $C_1 R_1$.

An expression for the pulse duration $t_p$ may be obtained by reference to Fig. 7.14. Figure 7.14(a) shows the voltage waveforms for $TR_2$. The trigger pulse switches on $TR_1$ causing the base of $TR_2$ to go suddenly negative at instant $A$, this being due to the charge on $C_1$. As $C_1$ discharges, this negative voltage decreases until eventually, at instant $B$, $TR_2$ is allowed to conduct again and the stable state is resumed. In order to simplify the analysis we shall neglect the collector-emitter saturation p.d. (0·1 V) and the forward base-emitter p.d. (0·6 V) as applicable for the two transistors. It is reasonable to neglect these potential differences as they are small compared with the supply voltage. This gives the simplified diagram of Fig. 7.14(b). At $A'$ (equivalent to $A$ in Fig. 7.14(a)), $C_1$ is fully charged with its right-hand plate, and hence the base of $TR_2$, at $-10$ V. $C_1$ now proceeds to discharge through $R_1$. Notice that $R_1$ is connected to the $+10$ V rail and therefore

## Waveform Generators

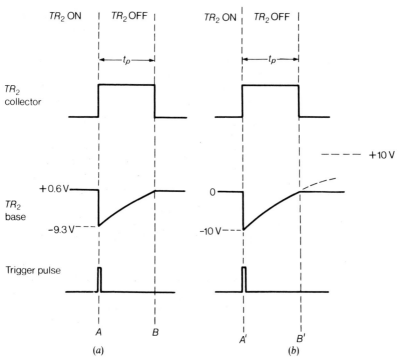

Fig 7.14 Voltages on $TR_2$ (a) as described in text (b) with simplifying assumptions.

$C_1$ will attempt to discharge and then recharge to $+10$ V. That is, the aiming voltage for $C_1$ is 20 V ($-10$ V to $+10$ V). $TR_2$ switches on as its base goes positive, i.e. after $C_1$ has charged to half the aiming voltage. Therefore, using equation (1), and putting the aiming voltage $V = 20$ V, and the voltage at which $TR_2$ conducts, $v = 10$ V, we get:

$$v = V(1 - e^{\frac{-t}{CR}}) \quad \text{where } t = t_p, C = C_1, R = R_1$$

$$10 = 20(1 - e^{\frac{-t}{CR}})$$

$$e^{\frac{-t}{CR}} = 0.5$$

$$e^{\frac{t}{CR}} = 2$$

$$\frac{t}{CR} = \ln 2$$

$$t = CR \ln 2$$

∴ $$t \simeq 0.7\, CR$$

Hence $$t_p \simeq 0.7\, C_1 R_1 \qquad (2)$$

Applying this result with $R_1 = 100$ k$\Omega$ and $C_1 = 5$ nF, the output pulse duration is:

$$t_p = 0{\cdot}7\, C_1 R_1$$
$$= 0{\cdot}7 \times 5 \times 10^{-9} \times 100 \times 10^3 \text{ s}$$
$$= 350\ \mu\text{s}$$

From equation (2), it is seen that the pulse duration $t_p$ is proportional to $C_1$ and $R_1$ and so it may be set to any desired time by changing the value of either component. Note, $R_1$ also determines the base current of $TR_2$ in the stable condition and therefore only a limited range of resistance values may be used. However, if $R_1$ is made up of a fixed and a variable resistor as in Fig. 7.15, small adjustments of $t_p$ may be made. When larger changes of pulse duration are required, the capacitor $C_1$ may be changed. The output pulse is taken from the collector of $TR_2$. When $TR_2$ is on, its collector voltage will be virtually zero. When $TR_2$ is off, its collector voltage will rise to a maximum value determined by the resistances $R_3$ and $R_4$.

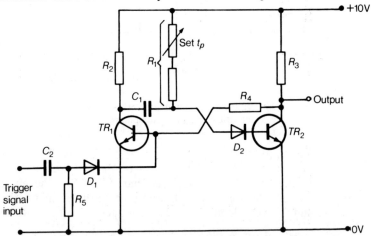

Fig 7.15 Monostable multivibrator with diode clipper at the trigger input.

The multivibrator shown in Fig. 7.15 is triggered by short duration positive pulses derived from the input signal by the clipper circuit involving $D_1$. The negative pulses produced by the differentiator are blocked by the diode as explained in section 7.4. The multivibrator sees only these short positive spikes and the input signal itself may be shorter or longer than the duration of the output pulse.

It was indicated earlier that the base of $TR_2$ goes negative by about 9·4 V. The maximum permissible reverse base-emitter voltage for silicon transistors however, is frequently only 5 V or 6 V. The diode $D_2$ helps to overcome this difficulty. In the stable state, the base current

# Waveform Generators

for $TR_2$ is supplied via this diode which is forward biased and its effect is insignificant. In the quasi-stable state, $D_2$ and the base emitter junction of $TR_2$ are both reverse biased. The reverse bias voltage is now shared between the diode and the base-emitter junction so that the reverse voltage rating of the latter is not exceeded.

## 7.6 Astable multivibrator

The astable (free running) multivibrator is a simple relaxation oscillator. A typical circuit is illustrated in Fig. 7.16. The two halves of the circuit are identical and so this particular oscillator will produce a square wave signal.

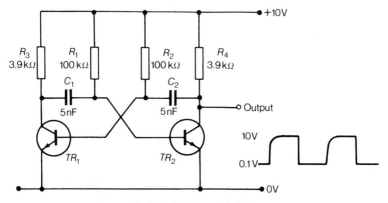

Fig 7.16 Astable multivibrator.

The circuit may be regarded as two simple amplifiers connected in a closed loop. If a small signal were to be applied at $TR_1$ base, it would be amplified and coupled to $TR_2$ base via $C_1$. $TR_2$ will then amplify the signal further and return it to $TR_1$ base via $C_2$. The two transistors are both operating as common emitter amplifiers and consequently each one inverts the signal. The signal returned to $TR_1$ base is therefore in phase with the initial signal. The loop gain for this circuit can be very high and so the necessary conditions for oscillation exist. When oscillation begins there is nothing to restrict the amplitude of the signal that is produced until the transistors are driven to either cut-off or saturation. Consequently, since the transistors are operating in anti-phase, one will be turned off, and the other turned on, alternately.

The connection of the supply will initiate this action and, almost immediately, one transistor turns off and the other turns on (assume $TR_1$ off and $TR_2$ on). The circuit is then in a quasi-stable condition like that described for the monostable multivibrator. After a time, the circuit will switch abruptly to a similar quasi-stable condition with $TR_1$ on and $TR_2$ off. The circuit continues to switch between these two states producing a square wave output signal in the process.

Having now recognised that a positive feedback action exists during the transition from one state to the other, it is much more realistic to consider the circuit operation in terms of the two quasi-stable conditions. It is assumed, in the description that follows, that the reader has studied section 7.5.

Let us start at a time immediately after $TR_1$ has turned on. The conditions in the circuit are as shown in Fig. 7.17. In the period prior to the changeover into this state, $C_1$ will have charged to 9·4 V. When $TR_1$ turns on, the base of $TR_2$ is driven negative and so $TR_2$ is turned off. Compare this situation with that illustrated in Fig. 7.13.

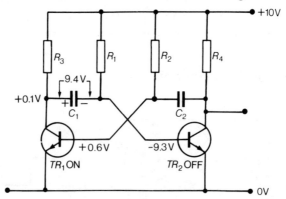

Fig 7.17 Conditions immediately $TR_1$ switches on.

When $TR_2$ switches off, its collector can rise to $+10$ V. This will not happen instantaneously because $C_2$ must charge via $R_4$ and the base emitter of $TR_1$ (the charging current causing a p.d. across $R_4$). As it does so, the collector voltage of $TR_2$ will rise. This accounts for the curved rising edge on the output signal waveform shown in Fig. 7.16. Eventually the collector of $TR_2$ reaches $+10$ V and, since the base of $TR_1$ is at $+0.6$ V, the p.d. across $C_2$ is 9·4 V. The circuit rests in this condition until $C_1$ is able to discharge via $R_1$ and $TR_1$, allowing the base of $TR_2$ to go positive. $TR_2$ will then start to conduct, and the feedback action indicated earlier will result in $TR_2$ being driven rapidly into saturation. The collector voltage of $TR_2$ falls abruptly to say 0·1 V, and as a result (because $C_2$ is charged), the base of $TR_1$ is driven negative. $TR_1$ therefore turns off. The conditions are now similar to those illustrated in Fig. 7.17 except this time $TR_1$ is off and $TR_2$ on. The collector voltage of $TR_1$ rises to $+10$ V as $C_1$ charges. The circuit then rests in this condition until $C_2$ discharges allowing $TR_1$ to conduct again. The whole process is then repeated and continuous oscillation results.

Using the analysis from section 7.5 it is clear that the period for which $TR_1$ is off is given by:

$$t_1 = 0.7\, C_2 R_2$$

# Waveform Generators

Similarly, the period for which $TR_2$ is off is given by:

$$t_2 = 0.7\, C_1 R_1$$

The period of one full cycle of the square wave output signal is therefore:

$$T = t_1 + t_2$$
$$= 0.7\, C_1 R_1 + 0.7\, C_2 R_2$$

Since the circuit in Fig. 7.16 is symmetrical (i.e. $C_1 R_1 = C_2 R_2$) the period in this case is:

$$T = 0.7 CR + 0.7 CR$$
$$= 1.4 CR \quad \text{where } C = C_1 = C_2$$
$$R = R_1 = R_2$$

The frequency of operation is therefore:

$$f = \frac{1}{T}$$
$$= \frac{1}{1.4\, CR} \text{ Hz}$$

The frequency of oscillation for the circuit in Fig. 7.16 with $R_1 = R_2 = 100$ kΩ and $C_1 = C_2 = 5$ nF is then:

$$f = \frac{1}{1.4\, CR}$$
$$= \frac{1}{1.4 \times 5 \times 10^{-9} \times 100 \times 10^3} \text{ Hz}$$
$$= 1428 \text{ Hz}$$

The operation of an astable multivibrator is summarised by the waveforms in Fig. 7.18. The astable multivibrator in this case is asymmetrical with $C_1 R_1 = 4 C_2 R_2$.

The astable multivibrator may be used to produce a sawtooth output waveform by using the principle outlined in section 7.3. A suitable circuit is illustrated in Fig. 7.19. Whilst $TR_2$ is off, $C_3$ will charge via $R_4$ and the output voltage will rise towards $+100$ V. When $TR_2$ conducts, it will discharge $C_3$ and reduce the output voltage to virtually 0 V. When $TR_2$ turns off again, $C_3$ will begin to charge towards $+100$ V again, and so on. The astable multivibrator must be asymmetrical with $TR_2$ conducting for a short time to discharge $C_3$ and then turning off for a much longer time to allow $C_3$ to recharge (i.e. $C_1 R_1 > C_2 R_2$). The output waveform will be a reasonably linear ramp provided $C_3$ is allowed to charge to only a small fraction of its aiming voltage. In

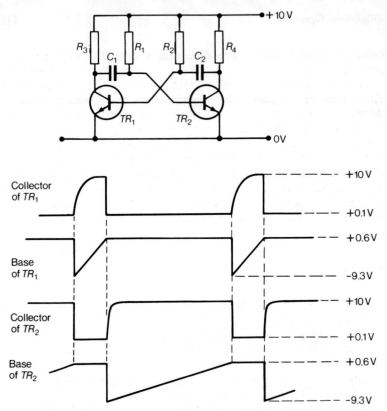

Fig 7.18 Waveforms for an asymmetrical astable multivibrator with $C_1R_1 = 4C_2R_2$.

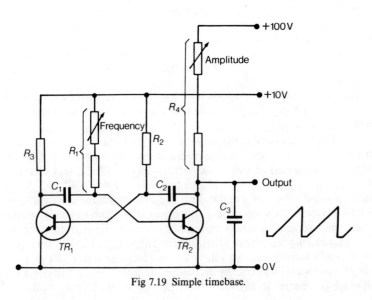

Fig 7.19 Simple timebase.

# Waveform Generators

order to obtain a linear ramp of say 20 V max., the aiming voltage needs to be substantially greater than this, such as the 100 V used here.

The frequency of the sawtooth waveform may be varied by adjusting $R_1$. For example, reducing the resistance of $R_1$ will reduce the time for which $TR_2$ is off and hence will increase the frequency of the output signal. Reducing the time for which $TR_2$ is off, will also reduce the amplitude of the sawtooth output signal because $C_3$ has less time in which to acquire charge. This may be compensated, if desired, by making $R_4$ adjustable. If the resistance of $R_4$ is reduced, $C_3$ will charge more rapidly. The output will therefore be able to rise to a higher voltage in the period for which $TR_2$ is off. Likewise, the amplitude of the sawtooth output signal can be reduced by increasing the resistance of $R_4$.

## Problems
(Answers at end of book)

1. Determine the time constant for a $C$-$R$ network in which $R = 120\,\text{k}\Omega$ and $C = 4\cdot7\,\text{nF}$.
2. Plot a charging curve for the p.d. across the capacitor in the circuit of Fig. 7.1(b) if $V_S = 150\,\text{V}$, $C = 10\,\text{nF}$ and $R = 100\,\text{k}\Omega$. Calibrate the horizontal axis in terms of both $\tau$ and time in ms.
   At what time is $v_c$ equal to 94·8 V, 130 V, 142·5 V? Deduce from the curve the time taken for $v_c$ to rise from 0 V to 75 V.
3. The network in question 2 is to be used to produce a voltage ramp with an amplitude of 50 V, from a 150 V supply, by the method outlined in section 7.3. Deduce the interval for which $S_1$ must be left open and hence the frequency of the sawtooth signal.
4. Given an integrator with $R = 50\,\text{k}\Omega$ and $C = 0\cdot01\,\mu\text{F}$ and an input signal as in Fig. 7.5, sketch the output voltage waveform.
5. Positive pulses with a duration of 5 ms, separated by intervals of 5 ms and with an amplitude of 5 V are applied to a differentiator in which $C = 50\,\text{nF}$ and $R = 20\,\text{k}\Omega$. Sketch the output waveform that will be obtained.
6. Draw a diode clipper circuit which will separate pulses corresponding to the negative going edge of a rectangular input pulse.
7. Given that $R_1$ in the monostable multivibrator in Fig. 7.11(a) is 82 k$\Omega$, calculate the value of $C_1$ such that a 115 $\mu$s output pulse is obtained.
8. Given that $R_1$ and $R_2$ in the astable multivibrator in Fig. 7.16 are both 100 k$\Omega$, calculate the capacitances of $C_1$ and $C_2$ such that the output waveform from $TR_2$ collector is a succession of positive pulses 140 $\mu$s duration, separated by intervals of 420 $\mu$s.

# Chapter 8
# Thermionic Valves

**8.1 Thermionic valves**
Many of the functions performed by semiconductor devices can also be fulfilled by thermionic valves, but the latter have significant disadvantages. Thermionic valves are bulky and mechanically fragile but, perhaps more significantly, they require a heater supply which makes them very inefficient. Compared to semiconductor equipment, valve equipment in general is less reliable, physically much larger, operates at higher potentials and temperatures, and requires better ventilation. The power supplies for such equipment are inevitably large and heavy. However, thermionic valves do have some features which are occasionally of value. They are well suited to high power applications and have an exceedingly high reverse resistance, and unlike semiconductor devices, they are not easily damaged by excessive forward or reverse voltage transients.

**8.2 Electron emission**
In Chapter 1, it was suggested that the electrons of an atom move further away from the nucleus as their energy is increased. When given the necessary energy, an electron can break away from its atom but it is still confined to the piece of material of which the atom is a part. However, if the material is enclosed in a vacuum, or low pressure gas atmosphere, electrons near the surface which have sufficiently high velocities may escape from the material. The release of electrons in this way is called electron emission. The energy required to cause the electron emission may come from a number of sources. Thermionic emission is of particular relevance to this chapter, but two other sources of electron emission are mentioned in passing.

*Thermionic Emission:* When a solid is heated, the electrons in the material acquire higher energies. At sufficiently high temperatures, electrons near the surface may attain velocities high enough to enable them to overcome the atomic forces that normally restrain them, and escape from the material. The release of electrons in this way is called *thermionic emission* and this is utilised in thermionic valves.

# Thermionic Valves

*Photo-Emission:* Photo-emission is electron emission which occurs when certain materials are exposed to electromagnetic radiation (e.g. visible light). The energy required to cause emission comes directly from the incident radiation. Photo-emission is utilised in certain types of photo-electric cell.

*Secondary Emission:* Secondary emission is electron emission which results from the bombardment of atoms by free electrons. For example, electrons released by thermionic emission may be accelerated towards a surface so that they strike it with a high velocity. On impact, these electrons give up their kinetic energy to the atoms at the surface. As a result, each incident electron may cause the release of several secondary electrons. This process is called secondary emission.

## 8.3 Diode valve

The construction of a directly heated thermionic *diode* valve is represented in Fig. 8.1(a) and its circuit symbol is given in Fig. 8.1(b). The device is called a diode because it has two electrodes, the *anode* and the *filament*. A current can be established between these electrodes by means of thermionic emission from the filament. The filament is heated directly by passing an electric current through it and so the valve is described as a *directly heated* thermionic diode. Connections to the electrodes are made via metal pins through the base of the valve (not shown in the diagram).

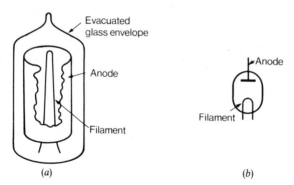

Fig 8.1 (a) Directly heated diode valve (anode shown cut away to reveal the filament). (b) Circuit symbol.

The anode is a metal cylinder, commonly made of nickel or, in larger valves, blackened steel. The filament is a thin tungsten wire which is suspended in the centre of the anode cylinder on insulating spacers. These spacers, which are typically made of mica, ensure electrical isolation of the two electrodes. The whole assembly is enclosed in a glass envelope and, during manufacture, all the air is pumped out to create a vacuum inside.

The filament is directly heated by passing an electric current through it. A pure tungsten filament is operated at around 2700 K (white heat). A small quantity of thorium oxide may be introduced into the tungsten (thoriated tungsten), in which case, the operating temperature is less at around 1800 K (yellow heat). At the operating temperature, substantial thermionic emission of electrons will occur. Electrons are continually released from the surface of the filament but, because the atoms they leave behind have a net positive charge, they will eventually fall back to it again. Nevertheless, a cloud of electrons will be produced around the filament which is continuously replenished by further electron emission. This is called a space charge cloud, or *space charge*. If the anode and the filament are deliberately connected together externally, (so that there is no p.d. between them) a small anode current is established. This current may be only a few microamperes, but it indicates that some electrons are thrown off the filament with sufficient energy to reach the anode (being returned to the filament via the external connection). If the anode is made negative with respect to the filament, even this very small current is stopped because electrons are now repelled from the anode. The reverse resistance of the valve, i.e. the resistance when the anode is made negative with respect to the filament, is therefore exceedingly high. On the other hand, if the anode is made positive with respect to the filament, the anode will attract electrons to it. Under these conditions, a substantial current can be established between the two electrodes. The d.c. resistance of the valve in the forward direction, i.e. the resistance when the anode is positive with respect to the filament, is therefore quite low. The electrons which make up this current are drawn from the space charge cloud surrounding the filament, rather than from the filament directly. Note, the electron movement from filament to anode constitutes a conventional current in the opposite direction.

## 8.4 Characteristics of diode valve

Figure 8.2 shows a circuit arrangement which can be used to determine the $I$-$V$ characteristics of a diode valve. The anode voltage can be

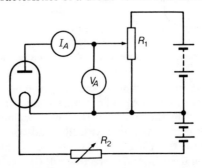

Fig 8.2 Circuit for determining $I$–$V$ characteristics of a diode valve.

## Thermionic Valves

varied by means of $R_1$ so that the relationship between the p.d. across the diode $V_A$, and the resultant anode current $I_A$, can be determined at a given filament temperature. Also, the filament current is made variable by means of $R_2$ so that the effect of changing the filament temperature can be examined. Typical $I$-$V$ characteristics are illustrated in Fig. 8.3. The forward characteristic has two distinct regions, called the *space charge limited* and *temperature limited* regions. Valves are normally operated in the space charge limited condition. It is not necessary to consider a reverse characteristic since there is no current when the anode is made negative.

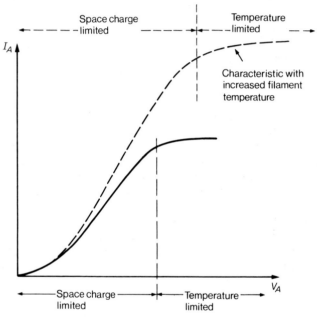

Fig 8.3 Static characteristics for a diode valve showing the effect of increasing the filament temperature.

Consider first the space charge limited region. As indicated earlier, the electrons emitted from the filament form a cloud around it. Electrons are, of course, negative charges and so this is a negative space charge cloud. This negative space charge tends to repel other thermally emitted electrons back to the filament. However, when the anode is made positive with respect to the filament, it attracts electrons towards it. At low anode voltages the effect of the space charge is dominant and there is only a small anode current. Those electrons which are gathered by the anode are drawn largely from the outer region of the space charge. As the anode voltage is increased, more and more electrons are drawn from the space charge cloud and the effect of the space charge is diminished. The increase in anode current for a given increase in anode voltage is therefore progressively greater as the anode voltage is raised.

The situation is eventually reached where the anode voltage is high enough to attract electrons directly from the filament as they are released by thermionic emission. That is, virtually all the electrons released from the filament are gathered immediately by the anode. Increasing the anode voltage further will not now cause a further increase in anode current because it is limited to the rate at which the electrons are emitted. The anode current is now said to have reached saturation level. The only way this saturation current can be increased is by raising the filament temperature thereby increasing the rate of supply of electrons. The effect of increasing the filament temperature is shown by the dotted curve in Fig. 8.3. Since it is the filament temperature that limits the anode current in this way, the saturation region of the characteristic is called the temperature limited region.

## 8.5 Indirectly heated diode valve

Tungsten and thoriated tungsten filaments are robust, but must be operated at high temperatures, and so require comparatively high power. Certain oxides will provide substantial thermionic emission at around 1000 K (red heat), but cannot conveniently be made into fila-

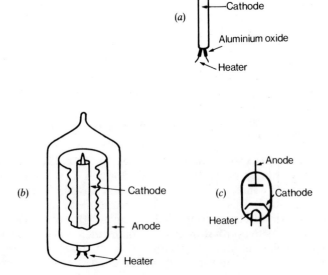

Fig 8.4 (a) Indirectly heated cathode.
(b) Indirectly heated diode valve.
(c) Circuit symbol.

# Thermionic Valves

ments. However, these oxides can be coated onto a thin nickel filament and, in effect, still be directly heated by passing a current through the filament itself. A more common practice is to use indirect heating as illustrated in Fig. 8.4(a).

The emitting material, typically a mixture of materials such as oxides of barium and strontium, is deposited onto a nickel tube. This forms the emitting electrode which is called the valve *cathode*. This tube is heated (to red heat) by a tungsten wire which passes through its centre. The tungsten wire, now called a *heater*, is electrically insulated from the cathode by a coating of aluminium oxide. As the thermionic emission occurs at lower temperatures, consideralby less heater power is required compared with directly heated tungsten filaments, and this form of construction is used for most thermionic valves. An indirectly heated diode valve is illustrated in Fig. 8.4(b) and the corresponding circuit symbol is shown in Fig. 8.4(c).

The principle of operation and the characteristics of indirectly heated valves are similar to those of directly heated types. The previous discussion of thermionic diodes therefore also applies to indirectly heated diodes.

## 8.6 The triode valve

The triode valve, as the name indicates, has essentially three electrodes. These are a cathode, an anode and a third electrode called the *control grid*. A heater is of course needed in an indirectly heated triode. The anode and cathode are essentially the same as for a diode valve. The control grid is an open fine wire structure, fitted between these two electrodes. The grid is able to influence the flow of electrons between the cathode and anode, and so makes it possible to use the triode valve as an amplifier. The arrangement is illustrated in Fig. 8.5(a) and the corresponding circuit symbol is shown in Fig. 8.5(b).

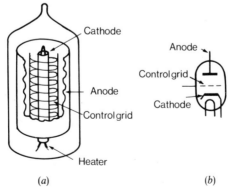

Fig 8.5 (a) Indirectly heated triode valve.
(b) Circuit symbol.

In normal operation, the anode is made positive with respect to the cathode which, ignoring the effect of the control grid, would normally cause an anode current. The electrons from the cathode must of course pass between the grid wires to get to the anode. Again, in normal operation, the control grid is made negative with respect to the cathode. This tends to repel electrons back towards the cathode and so restricts the anode current. Changes of grid to cathode voltage cause corresponding changes in anode current and, if the grid is made sufficiently negative, the anode current can be cut off completely.

## 8.7 Basic triode valve amplifier

Figure 8.6 shows a simple triode valve class $A$ amplifier. For normal operation, the control grid must be made negative with respect to the cathode. This is called *biasing* the valve and the cell $V_{GK}$ represents the *grid bias* in this circuit. The magnitude of the grid bias voltage is chosen to give the desired anode current. In practice, it is inconvenient to use cells or batteries for this purpose and the grid bias is developed automatically (see section 8.11). This arrangement however, clearly shows the required conditions.

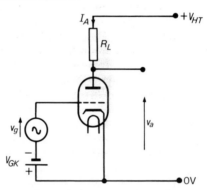

Fig 8.6 Basic triode valve amplifier.

The input signal $v_g$ is added to the grid bias which then represents the mean grid to cathode voltage about which the signal excursions occur. In normal operation, the peak value of the input signal must always be smaller than the grid bias voltage. Positive excursions of the input signal (which will make the grid less negative with respect to the cathode) will cause an increase in the anode current. Negative input signal excursions (which will make the grid more negative with respect to the cathode) will reduce the anode current. These changes of anode current in the load resistor $R_L$, will produce the output signal $v_a$. The amplitude of this output signal may be many times greater than the input signal, i.e. the stage can provide a substantial voltage gain.

Note, the cathode is common to the input and output circuits, and

# Thermionic Valves

so this is referred to as a common cathode amplifier. This is comparable with the common emitter amplifier in Chapter 5. A positive increase in the grid voltage will cause an increase in anode current. This will increase the p.d. across $R_L$ and so will reduce the anode voltage. Alternatively, when the grid goes more negative, the anode current will decrease and the anode voltage will rise. The output signal is therefore inverted relative to the input signal. Hence, like the common emitter amplifier, the common cathode amplifier is an inverting amplifier.

## 8.8 Static characteristics of triode valve

It is not necessary to consider an input characteristic for a triode valve. It has already been indicated that, in normal operation, the grid is negative with respect to cathode, and hence the grid current is negligible. The input resistance of the valve is therefore exceedingly high and may be disregarded.

The two useful sets of static characteristics are the output characteristics and the transfer (or mutual) characteristics. The static output characteristics for a low power triode valve are given in Fig. 8.7. These

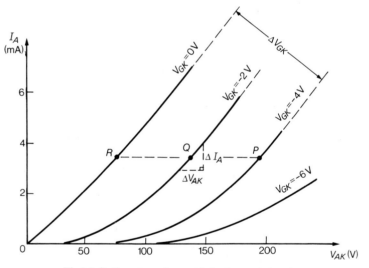

Fig 8.7 Static output characteristics for a triode valve.

show the relationship between the anode to cathode voltage $V_{AK}$, and the anode current $I_A$, for various values of grid to cathode voltage $V_{GK}$. The *output resistance* of the valve $r_a$ (often called the anode a.c. resistance) can be deduced from these characteristics. This may be defined as the ratio of a small change in anode voltage, to the corresponding small change in anode current, with the grid to cathode voltage being held constant.

i.e.
$$r_a = \frac{\Delta V_{AK}}{\Delta I_A}\bigg|_{V_{GK} \text{ constant}}$$

This is evaluated at point $Q$ in Fig. 8.7, as:
$$r_a = \frac{149 - 128}{(4 - 3) \times 10^{-3}}\bigg|_{V_{GK} = -2 \text{ V}}$$
$$= \frac{21}{1 \times 10^{-3}} \Omega$$
$$= 21 \text{ k}\Omega$$

The static transfer characteristics for the same triode valve are given in Fig. 8.8. These show the relationship between the anode current $I_A$, and the grid to cathode voltage $V_{GK}$, for given values of anode to cathode voltage $V_{AK}$. A parameter called the *transfer conductance* (or *mutual conductance*) $g_m$ can conveniently be deduced from these characteristics. This may be defined as the ratio of a small change in anode current, to the corresponding small change in grid to cathode voltage which causes it, with the anode to cathode voltage held constant:

i.e.
$$g_m = \frac{\Delta I_A}{\Delta V_{GK}}\bigg|_{V_{AK} \text{ constant}}$$

This is evaluated at the corresponding point $Q'$ in Fig. 8.8 as:
$$g_m = \frac{(4 - 2 \cdot 9) \times 10^{-3}}{-2 - (-2 \cdot 8)}\bigg|_{V_{AK} = 150 \text{ V}}$$
$$= \frac{1 \cdot 1 \times 10^{-3}}{0 \cdot 8} \text{ S}$$
$$= 1 \cdot 4 \text{ mS}$$

Note, this is change of current divided by change of voltage. The basic unit is therefore the Siemen, which is the unit of conductance. It has been common practice however, to quote mutual conductance in mA/V.

The third parameter of interest is the *amplification factor* $\mu$. It has been seen that a change in anode voltage will cause a change in anode current. Likewise, a change in grid voltage will change the anode current. The control grid however, has a much greater influence on the anode current than the anode itself. This is because the control grid is fitted between the anode and cathode, being quite close to the cathode surface. The amplification factor $\mu$, indicates the relative effectiveness of the anode and control grid in influencing the anode current. For example, assume that to obtain a certain change in anode current, the anode voltage must be changed by 5 V. It may be found that (for the same operating point) this same change in anode current could alternatively be produced by a change in grid to cathode voltage of only 0·1 V. The amplification factor is the ratio of these two changes, i.e. 5 to 0·1, or 50.

# Thermionic Valves

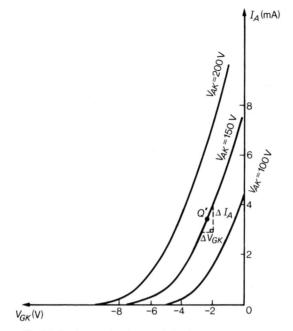

Fig 8.8 Static transfer characteristics for a triode valve.

The amplification factor may be defined as:

$$\mu = \left. \frac{-\Delta V_{AK}}{\Delta V_{GK}} \right|_{I_A \text{ constant}}$$

Note, making the anode more positive will increase the anode current. Similarly, making the grid more positive will increase the anode current. If however, there is to be no change of anode current (i.e. $I_A$ is constant), the changes of anode and grid voltage must be of opposite polarities. The above definition will therefore result in $\mu$ being a positive quantity. The amplification factor can be deduced from either set of static characteristics. Using the static output characteristics in Fig. 8.7, consider the points $P$ and $R$. These points for which $I_A$ is the same (ideally a smaller change of $V_{AK}$ and $V_{GK}$ would be used).

$$\begin{aligned}
\mu &= \left. -\frac{\Delta V_{AK}}{\Delta V_{GK}} \right|_{I_A \text{ constant}} \\
&= \left. \frac{-(195 - 75)}{(-4 - 0)} \right|_{I_A = 3.5 \text{ mA}} \\
&= \frac{-120}{-4} \\
&= 30
\end{aligned}$$

This is the ratio of two voltages and so is a dimensionless quantity (i.e. it is a ratio and has no units).

A circuit arrangement which may be used to determine the static characteristics for a triode valve is illustrated in Fig. 8.9. To determine the output characteristics, $R_1$ is set to give the desired value of $V_{GK}$, which is then held constant, whilst $V_{AK}$ is varied by means of $R_2$. Suitable values of $V_{AK}$ and the corresponding values of $I_A$ are recorded and plotted. To determine the transfer characteristics $R_2$ is set to give the desired value of $V_{AK}$ and then $V_{GK}$ is varied by means of $R_1$. Suitable values of $V_{GK}$ and the corresponding values of $I_A$ are then recorded and plotted.

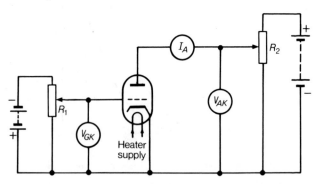

Fig 8.9 Circuit for determining static characteristics for a triode valve (heater supply typically a low voltage a.c. supply).

### 8.9 Relationship between $\mu$, $g_m$ and $r_a$

The static characteristics of a triode valve are to some extent non-linear. The actual values of the parameters $\mu$, $g_m$ and $r_a$ therefore depend on the operating point at which they are determined. However, for the same operating point, there is a simple relationship between the three which is demonstrated below.

A small change in anode current can be produced by a small change in anode voltage

i.e. $$\Delta I_A = \frac{\Delta V_{AK}}{r_a}$$

A small change in anode current can also be produced by a small change in grid voltage

i.e. $$\Delta I_A = g_m \Delta V_{GK}$$

Let the two changes be such that the anode current remains the same (remember that $\mu$ is defined with $I_A$ constant) then:

*Thermionic Valves* 125

$$\frac{\Delta V_{AK}}{r_a} + g_m \Delta V_{GK} = 0$$

∴ $$g_m \Delta V_{GK} = \frac{-\Delta V_{AK}}{r_a}$$

hence $$g_m r_a = \frac{-\Delta V_{AK}}{\Delta V_{GK}}$$

but $$\mu = \frac{-\Delta V_{AK}}{\Delta V_{GK}}$$

∴ $$\mu = g_m r_a$$

Substituting the values found in section 8.8.

$$(gm = 1\cdot 4 \text{ ms}, r_a = 21 \text{ k}\Omega)$$

$$\begin{aligned}\mu &= g_m r_a \\ &= 21 \times 10^3 \times 1\cdot 4 \times 10^{-3} \\ &= 29\cdot 4\end{aligned}$$

(which is nominally the value for $\mu$ which was deduced from the static output characteristics)

## 8.10 Voltage amplification

When a small signal voltage is applied at the control grid of the triode valve, it will cause corresponding variations of anode current. The transfer characteristic is reasonably linear over a small range, so provided the input signal swing is small, the anode current variations will be a replica of the input signal. These anode current variations in the load resistance $R_L$ of Fig. 8.6, will produce a corresponding output signal. The ratio of the output signal voltage to the input signal voltage is the voltage gain $A_v$.

i.e. $$A_v = \frac{v_a}{v_g}$$

Note, this is not the amplification factor $\mu$ which is of course defined for constant $I_A$.

Let us now consider an expression for the voltage gain for this type of triode valve amplifier. If, as seen from the anode, the valve appeared to be a perfect voltage generator (i.e. a voltage generator with zero output resistance), the voltage amplification would be equal to the amplification factor $\mu$ for the valve. That is, with an input signal of $v_g$ volts, the output signal would be $-\mu v_g$ volts. The minus sign represents the inversion of the signal which occurs with a common cathode amplifier. However, the valve does have an internal resistance, this being the output resistance $r_a$. Consequently, only a proportion of $-\mu v_g$ volts will appear at the output. This proportion is determined by the potential divider action of $R_L$ and $r_a$, and the actual output signal voltage will be:

$$v_a = -\mu v_g \times \frac{R_L}{r_a + R_L}$$

$$\therefore \quad \frac{v_a}{v_g} = \frac{-\mu R_L}{r_a + R_L}$$

but

$$\frac{v_a}{v_g} = A_v$$

hence

$$A_v = \frac{-\mu R_L}{r_a + R_L}$$

Also, since $\mu = g_m r_a$, this could equally be written as:

$$A_v = \frac{-g_m r_a R_L}{r_a + R_L}$$

These are general expressions for the voltage gain for a triode amplifier. We are now in a position to examine the practical triode amplifier circuit represented in Fig. 8.10.

### 8.11 Triode valve amplifier

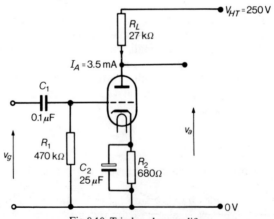

Fig 8.10 Triode valve amplifier.

It is normal practice to produce the grid bias voltage automatically by what is called automatic grid biasing or auto-cathode biasing. The requirement is to make the grid negative with respect to the cathode. This can be achieved by returning the control grid to the 0 V line, and making the cathode positive with respect to 0 V by an amount equal to the intended grid bias voltage. The grid is then negative with respect to

# Thermionic Valves

the cathode as required. Note that in the amplifier of Fig. 8.10 the grid is connected to 0 V via $R_1$. Since the grid current is negligible, there is no direct voltage across this resistor and so the grid is at 0 V. If the intended grid bias is $-2\cdot 4$ V, the cathode must be held at $+2\cdot 4$ V, and hence this is the p.d. required across $R_2$. Given that the anode current is to be 3·5 mA the resistance of $R_2$ is:

$$R_2 = \frac{V_{R_2}}{I_A}$$

$$= \frac{2\cdot 4}{3\cdot 5 \times 10^{-3}}\,\Omega$$

$$= 686\,\Omega \text{ (preferred value 680 }\Omega\text{)}$$

The values of anode current and grid bias used here have been chosen with reference to the static characteristics given earlier. That is, $I_A = 3\cdot 5$ mA and $V_{GK} = -2\cdot 4$ V correspond to point $Q'$ on the $V_{AK} = 150$ V characteristic in Fig. 8.8. This is a convenient operating

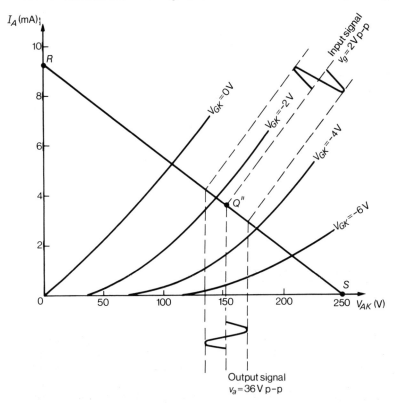

Fig 8.11 Load line for triode valve amplifier in Fig 8.10.

point, being in a region where the characteristics are reasonably linear. With $R_L = 27$ kΩ, and a supply voltage of 250 V, the anode voltage is:

$$\begin{aligned} V_A &= V_{HT} - V_{RL} \\ &= 250 - (3 \cdot 5 \times 10^{-3} \times 27 \times 10^3) \text{ V} \\ &= 156 \text{ V} \end{aligned}$$

The operating point can be related to the output characteristics by constructing a load line. This is shown in Fig. 8.11, and the operating point is point $Q''$. The co-ordinates of only two points are needed to position the load line, these being points $R$ and $S$. When the anode current is zero, there is no p.d. across the anode load resistor and so $V_{AK} = 250$ V (i.e. for point $S$, $V_{AK} = 250$ V, $I_A = 0$). Secondly, if $V_{AK}$ is zero, the anode current is approximately given by $V_{HT}$ divided by $R_L$ (the cathode resistor is ignored since it is small compared with $R_L$).

i.e.
$$\begin{aligned} I_A &= \frac{V_{HT}}{R_L} \\ &= \frac{250}{27 \times 10^3} \text{ A} \\ &= 9 \cdot 3 \text{ mA} \end{aligned}$$

(i.e. for point $R$, $V_{AK} = 0$, $I_A = 9 \cdot 3$ mA)

By constructing an input signal $v_g$ against the load line, the corresponding output signal $v_a$ may be determined as shown (see section 5.3 for further details of this technique). Taking peak-to-peak signal values from the diagram

$$\begin{aligned} A_V &= \frac{v_a}{v_g} \\ &= \frac{-36}{2} \\ &= -18 \end{aligned}$$

The operating point is nearly the same as that for which the valve parameters where deduced earlier (these were $\mu = 30$ and $r_a = 21$ kΩ). For the sake of comparison, we can calculate the voltage gain for the amplifier using the expression derived earlier

i.e.
$$\begin{aligned} A_V &= \frac{-\mu R_L}{r_a + R_L} \\ &= -\frac{30 \times 27}{21 + 27} \\ &= -17 \end{aligned}$$

# Thermionic Valves

This approximates to the value found by graphical methods as expected.

In Fig. 8.10, the capacitor $C_1$ is a coupling capacitor which will isolate the control grid from the d.c. conditions of the previous stage. It can be considered to be a short circuit to a.c. signals. The input signal is developed across the grid leak resistor $R_1$. When considering audio frequency signals, the input impedance of the valve may be neglected and the input impedance of the amplifier is then equal to $R_1$. The cathode decoupling capacitor $C_2$ short circuits $R_2$ as far as a.c. signals are concerned. This avoids the gain being reduced by the negative feedback action suggested in section 5.12.

## Problems
(Answers at end of book)

1. List the disadvantages of thermionic valves compared with semiconductor devices. Do the former have any significant advantages?
2. Given that a particular triode valve has an amplification factor of 55 and an output resistance of 26 kΩ, determine the mutual conductance for the same operating point.
3. Plot the static output characteristics for the triode valve for which the data below applies. Mark the operating $Q_1$ for which $V_{AK}$ = 120 V, $I_A$ = 2 mA.
   Determine $r_a$ and $\mu$ at this point.

|              | $I_A$ (mA) |        |         |         |         |
|--------------|------------|--------|---------|---------|---------|
| $V_{AK}$     | 40 V       | 80 V   | 120 V   | 160 V   | 200 V   |
| $V_{GK} = 0$   | 1·8        | 4·0    | 6·4     | —       | —       |
| $V_{GK} = -1$ V | 0·2        | 1·4    | 3·2     | 5·7     | —       |
| $V_{GK} = -2$ V | —          | 0·3    | 1·2     | 2·8     | 5·2     |
| $V_{GK} = -3$ V | —          | —      | 0·3     | 1·0     | 2·2     |

4. Using the output characteristics from question 3, and assuming a triode valve amplifier of the type shown in Fig. 8.10 but having a supply voltage of 200 V, construct a load line for a load resistor of 39 kΩ.
   Locate the operating point $Q_1$ on this load line.
   Assuming an input signal of 1 V p–p, determine the corresponding output signal voltage and hence deduce the voltage gain. Calculate the voltage gain, using the values for $\mu$ and $r_a$ deduced in question 3, and compare the result with that obtained by graphical means.

# Chapter 9
# The Cathode Ray Tube

**9.1 Principle of operation**

The cathode ray tube (c.r.t) is the basis of the cathode ray oscilloscope, the domestic television receiver and many other forms of visual display such as those used in radar and computer systems. A basic c.r.t. is illustrated in Fig. 9.1. It is essentially a funnel-shaped glass envelope in which a vacuum is created by pumping out the air. There is an electron gun at one end and a fluorescent screen at the other. The electron gun produces a thin pencil-like beam of electrons. This beam is directed at the screen, the inner surface of which is coated with a *phosphor* material which fluoresces (i.e. radiates visible light) when bombarded by the electrons in the beam. A spot of light is therefore produced.

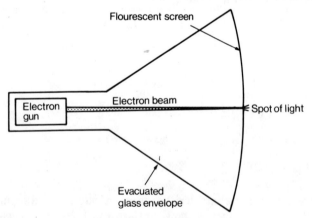

Fig 9.1 Basic cathode ray tube.

This spot may be deflected either horizontally or vertically, or in both directions simultaneously, by means of a so-called deflection system (not shown in the figure). The desired display, e.g. a television picture, is produced by moving the spot regularly over the screen and varying its brightness as required.

# The Cathode Ray Tube

## 9.2 Phosphor screen

The phosphor screen produces a visible light output when bombarded by the electrons in the beam. The colour of the light is determined by the particular phosphor material. The brightness (intensity) of the spot produced at the screen depends on the energy of the incident electron beam. Therefore, the brightness can be increased by increasing the velocity of the electrons in the beam (i.e. raising the *final anode* potential—see section 9.3), or by increasing the number of electrons which arrive at the screen in a given time (i.e. increasing the *beam current*). The latter is generally much more convenient.

## 9.3 Simple electron gun

A very basic electron gun is shown in Fig. 9.2. Electrons are released by thermionic emission from the indirectly heated cathode. They are then attracted through the grid and towards the screen by the anode which is held at a positive potential with respect to the cathode.

The cathode is in the form of a tube with the heater passing up the centre. The tube is capped at one end by a disc on which the emitting material is deposited. Adjacent to this emitting surface is the control grid. In Fig. 9.2 this is shown as a disc with a single small hole at its centre. In practice, it is commonly a cylinder with one end closed except for a single small hole in the centre of the closed end (as shown in Fig. 9.5). Electrons from the cathode are drawn through the hole in the control grid by the positive anode potential. Since there is only a single hole, the electrons form a beam with circular cross-section.

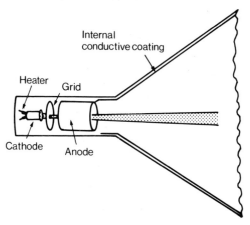

Fig 9.2 Simple electron gun.

In normal operation the control grid is made negative with respect to the cathode and the beam current can be varied by changing this grid to cathode potential. That is, the beam current and consequently the brightness of the spot can be increased by making the grid less negative with respect to the cathode and vice versa. The anode is a hollow cylinder and the electrons pass through it (as opposed to being collected directly by it) and go on to strike the screen. This action is discussed further in section 9.5.

The electrons in the beam tend to diverge and so, in a practical c.r.t., there must be a beam focusing system. Magnetic focusing has been used, but electrostatic focusing arrangements are now common. An electrostatic focusing lens, or *electron lens* as it is known (see section 9.5), can be formed by using two or more anode cylinders held at different potentials. The anode closest to the screen is normally at the highest potential and is called the final anode. It is connected to a conductive coating on the inner surface of the glass envelope (as illustrated in Fig. 9.2). This coating represents an extension of the final anode cylinder, and may be a graphite coating or, as is common in picture tubes, a very thin aluminium film. This aluminium layer* covers the inside of the flared portion of the envelope and also the inner surface of the phosphor screen. It is sufficiently thin to allow the electrons in the beam to penetrate through to the phosphor, but nevertheless forms a conductive coating over its surface providing the 'return path' for the electrons in the beam. The aluminium layer also acts as a mirror, reflecting light from the rear of the screen forwards through the face-plate, and hence gives a brighter display. Graphite coatings cannot be extended over the phosphor surface and so, when a graphite coating is used, electrons are returned from the screen to the final anode by secondary emission. Since, in this case, there is no conductive layer over the screen, charge differences may build up over the screen surface causing undesirable brightness differences.

The connections to the electrodes of the tube are normally made through pins at the base except when very high final anode potentials are used. For example, in picture tubes where the final anode potential may be up to 24 kV, the final anode connection is made through the side of flare to the internal conductive coating.

### 9.4 Motion of electron in electric field

Before considering the electron gun in more detail we must first examine the motion of a free electron in an electric field. Figure 9.3 represents two parallel metal plates in a vacuum. When a potential difference is applied to the plates an electric field is established between them. The lines representing the distribution of the field (i.e. lines of electric flux), indicate the direction in which a free positive charge would be urged to move. An electron, being a negative charge, would

---

* The aluminium layer also overcomes an effect known as ion burn which occurs in magnetically deflected cathode ray tubes, but this is beyond the scope of this book.

# The Cathode Ray Tube

experience a force in the opposite direction. If fringing* effects are ignored, the electric field will be uniform as shown. Notice the flux lines are straight and they are perpendicular at the metallic surfaces.

If we imagine an electron 'sitting' between the plates, with no sideways component of velocity, the electric force will cause it to move directly upwards towards the positive plate as shown. The electron will be accelerated progressively until it strikes the positive plate whereupon it gives up the kinetic energy that it has gained from the electric field.

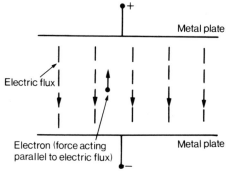

Fig 9.3 Force on an electron in an electric field.

Figure 9.4 shows the situation where an electron has been projected from an electron gun into the region between the plates. If there is no p.d. (i.e. no electric field) between the plates, the path of the electron will be unchanged and it will pass through in a straight line. When a p.d. with the polarity shown in the figure is applied, the electron will experience an upward force. Remembering that the electron has an initial velocity along the tube axis it is clear that it will be deflected upwards as shown at *A*. At a point such as *B* the electron is travelling in a new direction, but the upward force still applies, and so the electron is deflected further upward. Obviously, there is an upward force on the electron all the time it is in the electric field and so the path through the plates is a continuous curve as shown. When the electron leaves the region of the electric field, at *C*, it resumes a straight line trajectory in the direction now set. The amount by which the electron is deflected is dependent on its initial velocity as it enters the plates and the intensity of the electric field. The important thing to note for the moment however, is that the deflecting force acting at any point is always in line with the electric flux at that point, the direction being opposite to that marked on the flux lines (i.e. towards the more positive plate).

---

*The effect known as fringing describes the tendency of the flux lines at the edges of the plates to bulge out slightly.

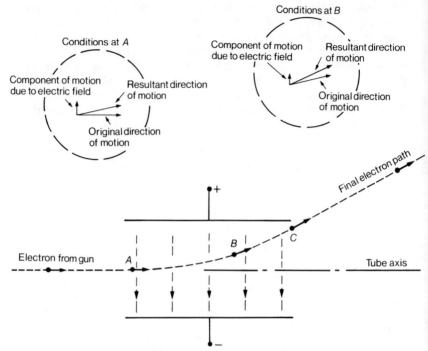

Fig 9.4 Deflection of an electron as it passes through an electrostatic field.

## 9.5 Electrostatic focusing

It was suggested earlier that the basic electron gun assembly will produce an electron beam with a circular cross-section. In fact, the control grid and its adjacent anode can be arranged to fulfil a further requirement. These electrodes are arranged so that the electric field between them, causes the electrons which pass through the control grid, to converge into a small area just outside the grid, called the *crossover* point. This is illustrated in Fig. 9.5. The cross-sectional area of the crossover is very much less than that of the cathode itself and this point represents the apparent origin of the electron beam. That is, it represents the 'object' which is focused at the screen by the main

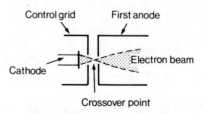

Fig 9.5 Formation of crossover.

# The Cathode Ray Tube

electron lens and so results in a smaller spot than would otherwise be achieved. The divergent beam from the crossover is focused at the screen by the electron lens which may involve two or three anode cylinders. A two anode lens is illustrated in Fig. 9.6. The two anodes are at different potentials and the electric field between them is as shown. The diagram represents a section through the anode cylinders and so, in fact, an electric field of this form completely surrounds the electron beam. At the centre of the cylinders (i.e. along the centre line), the electric flux is acting in the direction of the tube axis. An electron

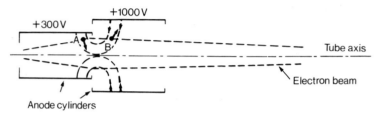

Fig 9.6 Focusing action of a two anode electron lens.

which enters from the left, travelling along the tube axis, will consequently be accelerated but not deflected. It therefore continues along the axis and strikes the centre of the screen. An electron which is moving away from the tube axis as it enters the left-hand anode must be brought back to the centre line by the time it reaches the screen. Consider one such electron when it reaches point $A$. Here it experiences a downward force in the direction shown. At $B$ the same electron will experience an upward force. The net deflection of the electron will be back towards the tube axis only if the downward deflection in the first region of the field is greater than the upward deflection in the latter region. At $A$, the downward component of force perpendicular to the tube axis, is greater than that at $B$. This is because, due to the different cylinder diameters, the direction of the electric force is closer to the perpendicular. Secondly, since the right-hand anode is at a higher potential, the electron will have been accelerated to a higher velocity when it passes through the region of field at $B$ and accordingly it is deflected less.

Since this electric field surrounds the electron beam, electrons diverging from the tube axis in any direction will be brought back to the focal point at the centre of the screen. Inspection of the electric field shows that the greater the divergence, the greater will be the forces acting on the electron to return it towards the tube axis. The focal length of the lens, and hence the focusing of the spot at the screen, may be adjusted by varying the relative potentials on the two anodes.

A complete gun assembly showing the crossover and beam focusing actions is illustrated in Fig. 9.7.

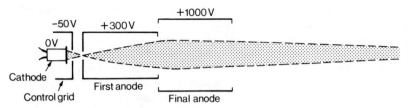

Fig 9.7 Basic gun assembly illustrating the crossover and focusing actions.

## 9.6 Electrostatic deflection

A c.r.t. of the type used in cathode ray oscilloscopes is shown in Fig. 9.8. This tube uses *electrostatic deflection*. This means that the electron beam, and hence the spot at the screen, is deflected by means of the electric field developed between metallic deflection plates. There are two pairs of deflection plates set at 90° to each other, one pair providing horizontal deflection and the other pair for vertical deflection.

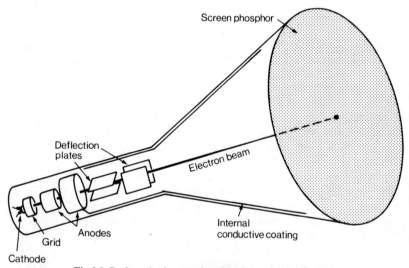

Fig 9.8 Basic cathode ray tube with electrostatic deflection.

The motion of an electron in an electric field was considered in section 9.4, but the situation is represented again in Fig. 9.9. It was shown that an electron projected through the plates is deflected towards the more positive plate. The path through the plates is a continuous curve which may be shown to be a parabola.

The deflection is found to be proportional to the potential difference between the plates. It is also proportional to the length of the plates $l$, and inversely proportional to the distance $d$, between them. In order to obtain reasonable deflection sensitivity (i.e. to avoid having to apply large deflection potentials to the plates) the plates must be placed quite

# The Cathode Ray Tube

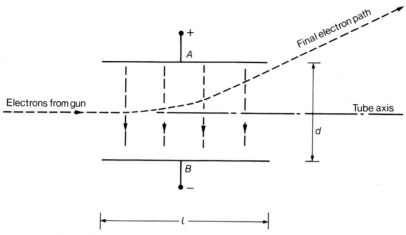

Fig 9.9 Electrostatic deflection. (*A* and *B* are metal deflection plates.)

close together, say 2 to 3 mm. Because of this, the two pairs of deflection plates must be placed in line (i.e. one pair ahead of the other) as in Fig. 9.8.

It is difficult to obtain large deflection angles because, if deflected too much, the beam will strike the plates. The plates could be made shorter or spaced further apart to avoid this problem, but both of these changes will reduce the deflection sensitivity. Commonly, the plates are splayed or the ends closest to the screen bent outwards to help overcome this difficulty.

Fig 9.10 Alternative configurations for deflection plates permitting larger deflection angles.

## 9.7 Electromagnetic deflection

Electromagnetic deflection is used for television picture tubes because much larger deflection angles can be achieved. This makes it possible to have a large screen on a relatively short c.r.t.

The principle of electromagnetic deflection is illustrated in Fig. 9.11 which shows how horizontal deflection is produced. Two coils are placed one above and one below the tube neck. When an electric current is passed through the coils, a magnetic field perpendicular to the tube axis is produced across the tube neck as shown. The magnetic circuit is completed externally by a ferrite yoke (not shown in the figure). The electron beam can be likened to a conductor carrying an

138  *Electronics for Technicians*

electric current and so, like such a conductor, there will be a force on the electron beam due to the magnetic field. Note that electrons moving towards the screen represent a conventional current in the opposite direction. The diagram therefore shows an electric current acting into

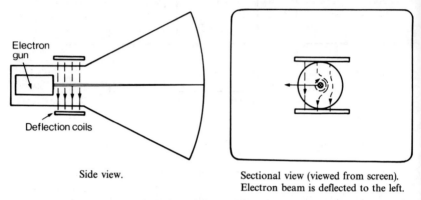

Side view.

Sectional view (viewed from screen).
Electron beam is deflected to the left.

Fig 9.11 Principle of electromagnetic deflection.

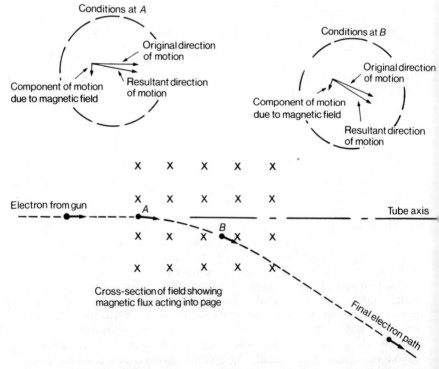

Fig 9.12 Deflection of an electron as it passes through a magnetic field.

# The Cathode Ray Tube

the page. The magnetic flux due to this current acts in a clockwise direction with the result that the flux density to the right of the beam is increased whilst that to the left is decreased. The electron beam is therefore deflected to the left as shown.

As indicated above, coils placed above and below the neck produce a vertical magnetic field and hence horizontal deflection of the beam. Similarly, coils placed either side of the neck will produce a horizontal magnetic field and hence vertical deflection of the beam.

The deflection of an electron as it passes through a magnetic field is illustrated in Fig. 9.12. The magnetic flux is shown acting into the page. Notice that the force acting on the electron is always at 90° to its direction of motion. The path of the electron through the magnetic field is therefore along the arc of a circle as shown. The deflection, for small deflection angles, is found to be approximately proportional to the magnetic flux density and hence to the current in the deflection coils.

Unlike the deflection plates considered earlier, the vertical and horizontal deflection coils can be placed in the same region along the tube neck. The coils can also be shaped to partly fit over the flared section of the tube which allows the tube neck to be shortened and permits larger deflection angles.

## 9.8 The cathode ray oscilloscope

Some further aspects of c.r.t. operation are now examined by considering the fundamentals of the cathode ray oscilloscope (c.r.o.). This instrument is commonly used to display electrical signal waveforms, but

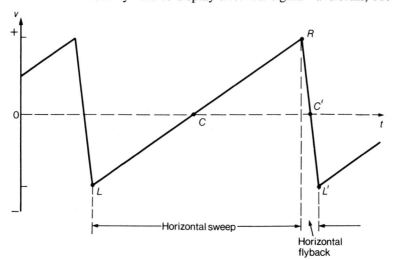

Fig 9.13 Sawtooth voltage waveform required to produce linear horizontal deflection of the electron beam in an electrostatically deflected c.r.t.

let us first consider how a single horizontal line can be produced on the screen.

An electrostatically deflected c.r.t. is used. The spot can be deflected horizontally across the screen by applying the sawtooth waveform, shown in Fig. 9.13, to the horizontal deflection plates ($X$ plates). This sawtooth waveform is produced by a so-called *horizontal timebase* generator. When the waveform is most negative (at $L$) the spot will be at the extreme left of the screen. As the potential difference between the

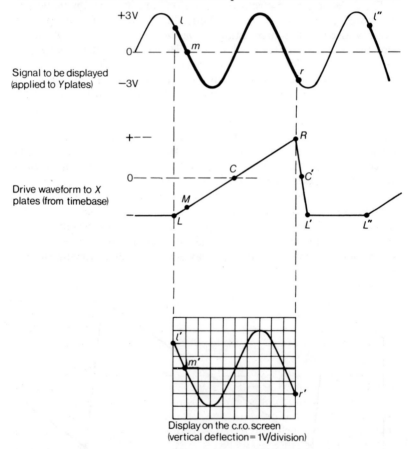

Fig 9.14 Display of a section of a sine wave signal on a c.r.o.

plates passes through zero (at $C$), the spot will move through the centre of the screen. At the positive maximum (at $R$) the spot is at the extreme right. Since the potential rises linearly from $L$ to $R$, the spot will move at a constant velocity from the left- to the right-hand side of the screen (remember, deflection is proportional to the p.d. between the plates).

# The Cathode Ray Tube

The movement from left to right is called the horizontal sweep. The more rapid change in potential ($R$–$C'$–$L'$) returns the spot quickly to the left-hand side again. This is called the horizontal flyback. At $L'$ the sweep begins again.

At low timebase speed settings the electron beam is deflected slowly and a clearly defined spot is seen to move across the screen. When deflected more rapidly this spot merges into a flickering line. When deflected even more rapidly a steady bright line is produced. This is because the screen phosphor has an *afterglow*. That is, the phosphor continues to emit light for a short time after it has been excited and the beam has moved on. This effect can be seen when the spot is moving slowly, the spot being followed by a comet-like tail of diminishing intensity. Another factor here is what is known as *persistence of vision*. This is a time-lag effect associated with the human visual process which is of particular significance in television and cine film operation.

The signal waveform that is to be displayed on the c.r.o. screen is applied, via an amplifier, to the vertical deflection plates ($Y$ plates) of the tube. A positive potential will deflect the beam upwards and a negative potential will deflect it downwards. Figure 9.14 shows how a selected section of a sine wave signal may be displayed. The horizontal sweep starts from the left (at $L$) when the sine wave signal has a magnitude of $+2$ V. The spot on the c.r.o. screen is therefore two divisions above the centre line at $l'$. Note, the vertical deflection is 1 V per division. At $M$, the sawtooth waveform on the $X$ plates has moved the spot horizontally one division, by which time the sine wave signal has fallen to zero, and so the spot is now on the horizontal centre line at $m'$. The horizontal and vertical deflections progress in this way until at $R$ the spot reaches the right-hand side with the $Y$ input signal at $-2$ V, at $r'$. At this point the horizontal flyback occurs. During flyback the spot does not retrace the previous path and so the beam is blanked. That is, the beam is shut off so that the flyback trace is not seen. This is called *flyback blanking* and suitable blanking pulses for the c.r.t. can be derived from the timebase generator.

A stable trace will be obtained only if an equivalent section of the $Y$ input signal is traced out during each horizontal sweep. That is, the next horizontal sweep must start at a point on the sine wave identical to that at which the previous one started. There is therefore a delay (period $L'$ to $L''$), before the next sweep is initiated (i.e. triggered off). This is called the *triggered mode* of operation, unlike the so-called *free running mode* suggested by Fig. 9.13, where the next sweep follows the flyback immediately.

## 9.9 Basic c.r.o.

Figure 9.15 shows a simplified diagram of a single beam cathode ray oscilloscope. The horizontal timebase generator produces the sawtooth waveform which is applied to the horizontal deflection plates ($X$

plates) via the $X$ amplifier. In the free running mode, the timebase produces a continuous sawtooth signal of the form shown in Fig. 9.13. The timebase speed is made adjustable and so it is possible, by careful adjustment, to obtain a stationary display in this mode. It will not remain stationary however, as this requires a constant frequency relationship between the timebase and $Y$ input signals. This difficulty is avoided in the triggered mode.

In the triggered mode, the sweep begins at a selected level on the input signal, the level being chosen by adjustment of the trigger level control. The sweep starts from an identical point on the $Y$ input signal each time. The duration of the section of $Y$ input signal that is displayed may be varied by adjusting the timebase speed. A trigger slope selector switch is also included. For the example shown in Fig. 9.14, the slope selector switch would be set to 'negative'. The sweep therefore starts at the selected level ($+2$ V) on the negative (downward) slope of the waveform. Alternatively, if the slope selector switch is set to 'positive', the sweep would start on the positive (upward) slope of the input signal.

The timebase speed setting is calibrated and so the time interval between points of interest may be measured. The frequency of a signal may therefore be determined by measuring its periodic time $T$, measured say as the period between two positive peaks, and calculating the frequency from $f = \dfrac{1}{T}$ Hz.

The signal to be displayed is applied to the vertical deflection plates ($Y$ plates) via the $Y$ amplifier. This amplifier is preceded by a variable input attenuator so that a wide range of input signal potentials may be displayed. The $Y$ amplifier has a fixed gain and the input attenuator reduces the $Y$ input signal to a level which gives a conveniently large display. This attenuator is calibrated in terms of the vertical deflection at the screen so that the magnitude of the $Y$ input signal may be measured.

It is common to drive the deflection plates in push–pull, i.e. with equal amplitude but opposite polarity signals to each plate as shown in Fig. 9.15. However, the principle of operation is still essentially as described.

With reference to the d.c. supplies, it is convenient to have the deflection plates at low potentials and so the final anode is grounded and the cathode is made nominally 1000 V negative with respect to earth (note the anode is still positive, and the grid still negative, with respect to the cathode). The potentials for the c.r.t. electrodes are tapped from a potential divider with the grid being the most negative electrode, as shown in Fig. 9.15. The setting of $R_1$ determines the grid to cathode potential. This sets the beam current and so acts as the intensity control. The beam focusing is adjusted by changing the focus

# The Cathode Ray Tube

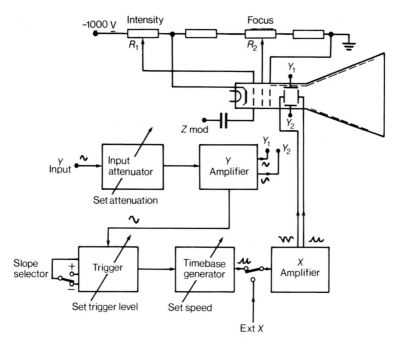

Fig. 9.15 Simplified diagram of a cathode ray oscilloscope.

anode potential (which in this simple tube is the first anode) by means of $R_2$.

Two other common facilities are shown. The c.r.o. may be used as an $X$-$Y$ plotter. In the $X$-$Y$ mode the internal timebase is disconnected and an external $X$ signal is applied to the $X$ amplifier.

A $Z$ modulation ($Z$ mod) input is provided so that the intensity of the spot may be modulated by varying the grid to cathode potential. The $Z$ mod signal is applied at the c.r.t. grid.

## Problems
(Answers at end of book)

1. State two reasons why the inner surface of a c.r.t. may have a very thin aluminium coating.
2. In an electrostatically deflected c.r.t., what would be the effect of changing each of the factors below (i.e. one at a time, the other factors remaining unchanged).
   (a) Increasing the length of the horizontal deflection plates;
   (b) Reducing the distance between the $Y$ plates;
   (c) Increasing the final anode voltage;
   (d) Making the grid less negative with respect to the cathode.

3. What is the nature of the curved path which electrons follow when passing through:
   (a) an electrostatic field?
   (b) a magnetic field?
4. The deflection coils on a magnetically deflected c.r.t. can be moved along the tube neck. What will be the effect of moving the deflection coils a small distance towards the gun end of the neck?
5. What will be the effect of applying a direct voltage to the $Y$ plates of an electrostatically deflected c.r.t. in a cathode ray oscilloscope?

# Chapter 10
# Logic Gates and Circuits

## 10.1 Logic systems

Logic systems are constructed using two-state devices. Figure 10.1 represents a very simple logic circuit in which the switch (open or closed) and lamp (on or off) are both two state devices. The circuit can be considered to have an input and an output. Operating switch $A$ represents the input, and energising lamp $F$ represents the output. The behaviour of the circuit may be summarised in what is known as a truth table which shows all the possible input combinations and the resultant output states. For this purpose, it is convenient to refer to the two states as the logic 0 and logic 1 states. For the input (switch $A$), 0 represents 'switch open' and 1 represents 'switch closed'. For the output (lamp $F$), 0 represents 'lamp off' and 1 represents 'lamp on'. This considerably simplifies the writing of a truth table as shown in Fig. 10.2.

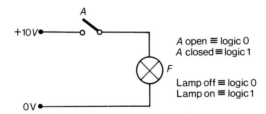

Fig 10.1 A simple two-state system ($\equiv$ means 'equivalent to').

| Input<br>A | Output<br>F |
|---|---|
| Switch open | Lamp off |
| Switch closed | Lamp on |

$\equiv$

| Input<br>A | Output<br>F |
|---|---|
| 0 | 0 |
| 1 | 1 |

Fig 10.2 Alternative truth tables for the circuit in Fig 10.1.

Logic 0 and logic 1 commonly represent direct voltage levels in an electrical circuit. In Fig. 10.1, the potential difference across the lamp will either be 0 V or 10 V. Using the convention already established for the lamp we get:

| P.D. across lamp | Output state |
|---|---|
| 0 V | logic 0 |
| +10 V | logic 1 |

The inputs may also be direct voltage levels. For example, if switch *A* is a pair of normally open contacts on a relay, the direct voltage at the input would be the p.d. applied to the relay coil. Assume the relay contacts (switch *A*) close when +10 V is applied to the relay coil. The logic definitions for the input are then:

| P.D. across relay coil | Input state |
|---|---|
| 0 V | logic 0 |
| +10 V | logic 1 |

A system like this, in which the more positive of two direct voltage levels is defined as the 1 state, is called a positive logic system. If the more negative of the two levels is assigned to the 1 state, the system is called a negative logic system. Discussion in this chapter is confined to positive logic systems only.

## 10.2 Communication by two-state systems

A two-state system might not appear to be very useful for the transmission of information down a single pair of wires, or by means of a single radio channel, because only two signal levels are used. By using predetermined codes however, it is possible to transmit complex messages. An obvious example is the use of Morse code. Two-state signals are employed (e.g. an audible tone or no tone). Tone bursts of different lengths are used to represent 'dot' and 'dash' respectively and, by suitable coding, each letter of the alphabet can be separately identified. Words are separated from each other by using a longer space.

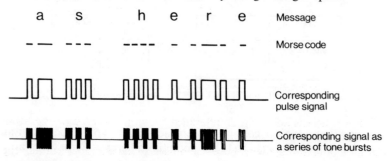

## Logic Gates and Circuits

Consider another example in which any number in the range 0 to 7 is to be transmitted. These numbers might represent commands to activate any one of eight possible functions (opening or closing control valves say) in a piece of equipment which is operated remotely via only one pair of wires. The information relating to a particular number may be sent in the form of a binary code as shown in Fig. 10.3. Three binary digits (abbreviated—bits) are required to represent the decimal numbers 0 to 7. This means there is a maximum of three pulses in each data interval. A 'pulse' here represents 1 and 'no pulse' represents 0. The diagram shows the transmission of four separate commands. For example, function 4 is represented by binary 100 and so is transmitted as 'pulse', 'no pulse', 'no pulse'. Similarly, function 5 is selected by sending the code 101 ('pulse', 'no pulse', 'pulse'), and so on.

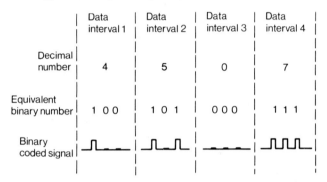

Fig 10.3 Example of a binary coded signal

A complete set of signals representing the eight possible commands are shown in Fig. 10.4.

| Decimal number | Equivalent binary number | Binary coded signal |
|---|---|---|
| 0 | 0 0 0 | |
| 1 | 0 0 1 | |
| 2 | 0 1 0 | |
| 3 | 0 1 1 | |
| 4 | 1 0 0 | |
| 5 | 1 0 1 | |
| 6 | 1 1 0 | |
| 7 | 1 1 1 | |

Fig 10.4 Binary coded signals for decimal numbers 0 to 7.

Notice there is an essential difference between this type of operation and the Morse code example. The pulses here are all of the same duration and the spacing between pulses is constant, or rather, the spacing between the intervals during which pulses may occur is constant. This type of operation could be extended to include a complete alphabet by using more complex codes. Complete written messages may then be transmitted.

## 10.3 Logic gates

Logic systems, in which these two-state signals are processed, are built up by interconnecting a number of basic logic gates, three of which are considered in this chapter. Each basic gate performs a simple, but clearly defined, logic function. By using many gates, and interconnecting them in different ways, it is possible to create very complex systems such as computers, guidance systems and sophisticated communications equipment. The logic function that a gate fulfills may be illustrated in a number of ways. The input and output states may be represented in a truth table. Alternatively, waveform diagrams may be drawn to show the direct voltage levels at the output and inputs and how they are related. These are called timing diagrams. Another method is to use a special form of algebra known as Boolean algebra.

## 10.4 The AND gate

The logical AND function is illustrated by the circuit of Fig. 10.5(a). The lamp will light only if switch $A$ AND switch $B$ are both closed. Obviously, the lamp will not light if only one switch is closed. The operation may be represented in a truth table using the notation estab-

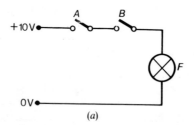

| Inputs | | Output |
|---|---|---|
| Switch $A$ | Switch $B$ | Lamp $F$ |
| Open | Open | Off |
| Open | Closed | Off |
| Closed | Open | Off |
| Closed | Closed | On |

| Inputs | | Output |
|---|---|---|
| $A$ | $B$ | $F=A.B$ |
| 0 | 0 | 0 |
| 0 | 1 | 0 |
| 1 | 0 | 0 |
| 1 | 1 | 1 |

(b)

Fig 10.5 (a) Representation of a 2-input AND gate.
(b) Truth table for a 2-input AND gate.

## Logic Gates and Circuits

lished earlier. As there are two switches, there are four possible input combinations, and so there are four rows in the truth table of Fig. 10.5(b).

In practice, logic gates can be made using discrete diodes and/or transistors, but they are readily available as integrated circuits and they are normally used in this form. The particular circuit arrangement for a gate is generally of no interest and so logic symbols, which show the logic function of the gate, are sufficient when drawing system diagrams. Three alternative symbols for an AND gate are shown in Fig. 10.6.

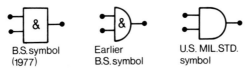

Fig 10.6 Alternative logic symbols for a 2-input AND gate.

In Fig. 10.7, the *AND* function is illustrated by direct voltage levels at the inputs and output of the gate. Notice that an output is obtained only when both inputs are high (at logic 1). The output stays low (at logic 0) if either, or both of the inputs are low.

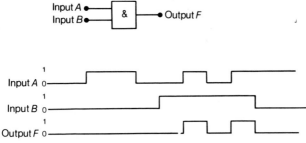

Fig 10.7 Operation of AND gate illustrated by voltage waveforms.

The logical AND function may also be represented in Boolean algebra as:

$$F = A.B$$

The [.] represents the AND function and this equation is read as:

$$F \text{ equals } A \text{ AND } B$$

### 10.5 The OR gate

The logical OR function is illustrated by the circuit of Fig. 10.8(a). The lamp will light if either switch *A* OR switch *B* is closed. The lamp will also light if both switches are closed and, for this reason, this is sometimes called the INCLUSIVE-OR function. The operation is represented in a truth table in Fig. 10.8(b).

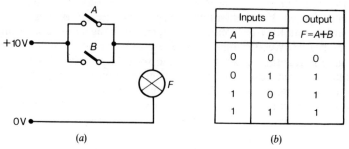

Fig 10.8 (a) Representation of a 2-input OR gate.
(b) Truth table for a 2-input OR gate.

The relevant logic symbols are shown in Fig. 10.9.

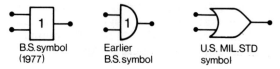

Fig 10.9 Alternative logic symbols for a 2-input OR gate.

The operation of the gate is shown in terms of direct voltage levels in Fig. 10.10. The input signals are the same as before, but, this time, an output is obtained if input $A$ OR input $B$ is high.

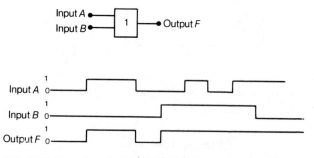

Fig 10.10 Operation of OR gate illustrated by voltage waveforms.

The logical OR function is represented in Boolean algebra as:

$$F = A + B$$

The [+] denotes the OR function and the equation is read as:

$F$ equals $A$ OR $B$

# Logic Gates and Circuits

## 10.6 The NOT gate

The NOT gate or INVERTOR is illustrated by the circuit of Fig. 10.11(a). The lamp is connected to a relay the contacts of which are normally closed. If switch $A$ (which represents the input) is closed, the relay is energised, the contacts $R_1$ open, and the lamp goes out. If switch $A$ is opened, the relay is de-energised, contacts $R_1$ close again, and the lamp is turned on. So, the lamp will NOT light if switch $A$ is operated, i.e. the lamp will NOT light if the input is applied. This is referred to as inversion or negation of the input. Clearly, if the input is logic 1 the output is logic 0, and vice versa, giving a very simple truth table.

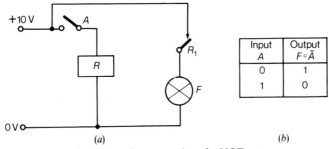

Fig 10.11 (a) Representation of a NOT gate.
(b) Truth table for a NOT gate.

The relevant logic symbols are shown in Fig. 10.12.

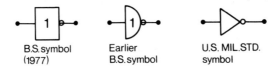

Fig 10.12 Alternative logic symbols for a NOT gate.

An invertor has only one input and the operation of the gate is illustrated in terms of direct voltage levels in Fig. 10.13. Notice the output is high when the input is low and vice versa.

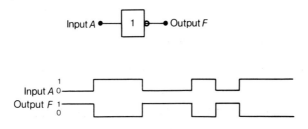

Fig 10.13 Operation of a NOT gate illustrated by voltage waveforms.

Inversion (or negation) is represented in Boolean algebra by putting a bar over the symbol. i.e. $\bar{A}$ means, and is read as, NOT $A$. Likewise, the equation that applies to Fig. 10.13 is:
$$F = \bar{A}$$
Note that if $A = 1$ then $\bar{A} = 0$ and vice versa.

## 10.7 Logic circuits

Consider a simple problem relating to a machine and the two conditions under which it may be permitted to operate.

These are:
1. The machine may operate if a work piece has been inserted and the guard is in position.
2. The machine may operate when there is no work piece if the maintenance engineer has operated the appropriate override switch and again the guard is in position.

There are then three input signals, so let:

$A$ represent the guard in position;
$B$ represent the work piece in position;
$C$ represent the override switch operated.

Let the output $F$ represent the output signal which must be present before the machine can be started.

The conditions for operation may be restated as follows:

The machine should operate only if the work piece is in position OR the override has been operated, AND the guard is in position. This is interpreted as a logic circuit in Fig. 10.14(a) and its corresponding truth table is given in Fig. 10.14(b). Since there are 3 inputs, there are 8 possible input combinations all of which are included in the truth table. It is convenient to write the possible input combinations in binary number order so that it is easy to check that all of them have been taken into account. In general, if there are n inputs, there will be $2^n$ possible input combinations, i.e. for 3 inputs, $2^n = 2^3 = 8$.

Column $D$ in the truth table, which represents the output of the OR gate, is completed first by reference to the $B$ and $C$ inputs, i.e. $D = B + C$. Then, by reference to the $A$ and $D$ columns, the final output is obtained as $F = A.D$, [i.e. the complete function is $F = A.(B + C)$]. A logic 1 appears in the output column for three rows in the table. This means there must be three sets of conditions which will allow the machine to operate.

Row (i)  The guard is in position, there is no work piece, but the override switch has been operated.
Row (ii)  The guard is in position, and the work piece is in position, but the override switch has not been operated.
Row (iii)  The guard is in position, the work piece is in position, but the override switch has been operated also (i.e. the INCLUSIVE-OR

condition for the B and C inputs applies). If this is unacceptable for some reason the circuit would have to be redesigned.

This simple example shows how logical statements may be interpreted into a logic circuit, and also, how the truth table can be used examine the operation of the circuit.

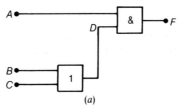

| Inputs  |   |   |           | Output        |       |
|---------|---|---|-----------|---------------|-------|
| A | B | C | D = B + C | F = A.(B + C) |       |
| 0 | 0 | 0 | 0 | 0 |       |
| 0 | 0 | 1 | 1 | 0 |       |
| 0 | 1 | 0 | 1 | 0 |       |
| 0 | 1 | 1 | 1 | 0 |       |
| 1 | 0 | 0 | 0 | 0 |       |
| 1 | 0 | 1 | 1 | 1 | (i)   |
| 1 | 1 | 0 | 1 | 1 | (ii)  |
| 1 | 1 | 1 | 1 | 1 | (iii) |

(b)

Fig 10.14 (a) Logic circuit for machine control system.
(b) Truth table for machine control system.

## 10.8 Function selector circuit

Now consider a different type of problem. A remotely controlled function selector system was suggested in section 10.2. In this system, each transmitted code represents a separate command to activate one out of eight possible functions. At the receiving end, there must be a decoding circuit which is able to recognise the transmitted code and produce a signal to activate the appropriate function. For simplicity, here we will consider a slightly less complicated system in which there are only four possible commands as set out below.

| Function Number | Equivalent Binary Number |   |
|---|---|---|
|   | A | B |
| 0 | 0 | 0 |
| 1 | 0 | 1 |
| 2 | 1 | 0 |
| 3 | 1 | 1 |

To activate a particular function, e.g. function 3 (open the water supply valve say), the appropriate binary signal must be received and held at the inputs of the decoder circuit as illustrated in Fig. 10.15; i.e. the $A$ and $B$ inputs applied simultaneously. With function 3 selected, both the $A$ and $B$ inputs are at logic 1. Gates $(a)$ and $(b)$ are invertors and so the $\bar{A}$ and $\bar{B}$ lines are at logic 0. Following the connections through to the inputs of the other gates, it is clear that gate $(f)$ has both of its inputs high and so its output is high also. This represents the command signal which causes the water supply valve to be turned on. Considering the other three gates, it can be seen that at least one of their inputs is low and so their outputs are all low. Try this operation again for function 2, i.e. put $A$ at 1 and $B$ at 0 and determine the input states for gates $(c)$ to $(f)$ as before. It should be found that the only gate which has both its inputs at logic 1 is gate $(e)$ and hence only function 2 is activated. The same principle can be applied to confirm the operation of gates $(c)$ and $(d)$.

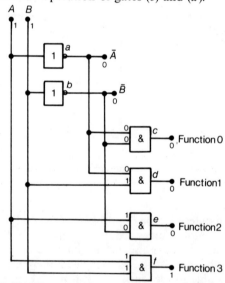

Fig 10.15 Function selector with function 3 selected.

## 10.9 Diode gates

Examples of the use of AND, OR and NOT gates have been considered. These gates may be made using discrete diodes and/or transistors but they are readily available as integrated circuits. For example, one integrated circuit may contain four 2-input AND gates, or six invertors, and so integrated circuits are usually the natural choice. The particular circuit arrangement for a gate is generally of no interest, but the principle of operation of the AND or OR gates are illustrated by the diode gate circuits which follow.

# Logic Gates and Circuits

## 10.10 Diode AND gate

Figure 10.16(a) shows the circuit diagram of a positive logic 3-input diode AND gate. Assume for convenience that the source resistances of the circuits feeding the inputs are zero, and the input levels are say +10 V (logic 1) and 0 V (logic 0). Having made these assumptions the operation of the circuit can be seen by reference to the single diode in Fig. 10.16(b). When $S_1$ is set to logic 0, the diode cathode is taken to 0 V, and the diode is forward biased via $R$ which is connected to a +10 V supply. Since, when the diode is conducting, the p.d. across it is say 0·6 V, the output potential is low (logic 0) being equal to the forward p.d. across the diode.

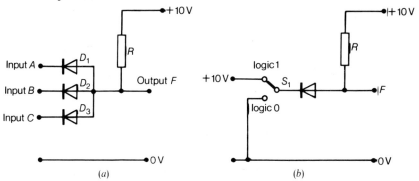

Fig 10.16 (a) 3-input positive logic diode AND gate.
(b) section of diode AND gate circuit considering one diode only.

When $S_1$ is set to logic 1, the diode cathode is at +10 V and, via $R$, its anode is returned to +10 V also. The diode is therefore unbiased, there is no current in the circuit, and so the output terminal is at +10 V (logic 1).

Now this can be extended to the AND gate circuit. If any one, or

| Inputs |   |   | Output |
|---|---|---|---|
| A | B | C | F = A.B.C |
| 0 | 0 | 0 | 0 |
| 0 | 0 | 1 | 0 |
| 0 | 1 | 0 | 0 |
| 0 | 1 | 1 | 0 |
| 1 | 0 | 0 | 0 |
| 1 | 0 | 1 | 0 |
| 1 | 1 | 0 | 0 |
| 1 | 1 | 1 | 1 |

Fig 10.17 Truth table for a 3-input AND gate.

more than one of the inputs is low, the output will be low also. For example, if input $A$ is low and inputs $B$ and $C$ are high, the output is held low because $D_1$ is conducting. Since they are connected together all three anodes are low, and as the cathodes of $D_2$ and $D_3$ are high, these two diodes are reverse biased. Only when all three inputs are high will the output go high and so the circuit satisfies the AND function. The truth table for the gate is given in Fig. 10.17.

## 10.11 Diode OR gate
Figure 10.18(a) shows the circuit diagram of a positive logic 3-input OR gate. Again assume the source resistances of the circuits feeding

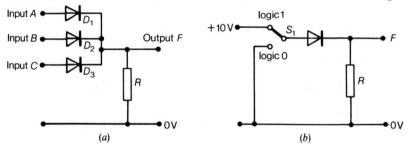

Fig 10.18 (a) 3-input positive logic diode OR gate.
(b) section of diode OR gate circuit considering one diode only.

the inputs are zero. Consider the circuit in Fig. 10.18(b) which has only one diode. The diode is forward biased when the input is high and so the output goes high also (equal to $+10$ V less the forward p.d. across the diode). When the input is low, i.e. 0 V here, there is no p.d. across the diode and no current in $R$. The output terminal is therefore at 0 V (logic 0).

Extending this to the 3-input OR gate it can be seen that, if all the inputs are low, the output stays low. If any one input goes high, the output will go high and the remaining diodes will merely be reverse biased. Obviously, if two or more inputs are high, the output is high also (INCLUSIVE-OR). The truth table for the circuit is given in Fig. 10.19.

## 10.12 The NOT gate
Figure 10.20(a) shows a NOT gate or invertor. The operation of this circuit is readily appreciated by considering the switching action of a transistor. When the input is low, the base-emitter junction of the transistor is reverse biased because $R_2$ is returned to a negative supply potential (i.e. the transistor is switched off). Hence, there is no collector current and consequently no p.d. across $R_3$. The output is therefore high (at $+10$ V).

When input $A$ is high, the transistor will be supplied with a base current, via $R_1$, sufficient to saturate it (i.e. switch it hard on). The

# Logic Gates and Circuits

| Inputs |   |   | Output |
|:-:|:-:|:-:|:-:|
| A | B | C | F = A + B + C |
| 0 | 0 | 0 | 0 |
| 0 | 0 | 1 | 1 |
| 0 | 1 | 0 | 1 |
| 0 | 1 | 1 | 1 |
| 1 | 0 | 0 | 1 |
| 1 | 0 | 1 | 1 |
| 1 | 1 | 0 | 1 |
| 1 | 1 | 1 | 1 |

Fig 10.19 Truth table for a 3-input OR gate.

output is then low being equal to the saturation potential of say 0·1 V at the collector. It follows that the output is always the opposite, or complement as it is known, of the input. This is as expected for an invertor.

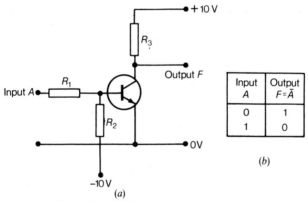

Fig 10.20 (a) NOT gate (invertor).
(b) Truth table for a NOT gate.

## Problems
(Answers at end of book)

1 Draw a logic circuit which will perform a function equivalent to the simple switching circuit in Fig. 10.21.

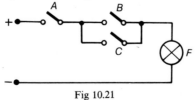

Fig 10.21

Construct a truth table for this circuit and write the corresponding Boolean expression.

2  Devise a function selector circuit like that in section 10.8, but providing selection of one out of eight functions. Check the circuit operation by applying appropriate input signals.

3  (a) A mixing tank has to be filled simultaneously with a mixture of three separate chemicals each from separate containers. The tank must be allowed to fill only when all three containers have reached predetermined temperatures. Assuming control signals are provided from each container to indicate when the temperature is correct, devise a simple logic system which will provide a common control signal for the outlet valves on the containers.

(b) Provide an alternative circuit assuming only 2-input gates are available.

4  (a) A laboratory has two entrances each of which has airlock type doors. At each entrance, one of the two doors should be closed before the other is opened. Devise a simple logic circuit, the output of which will operate an alarm in the event that the two doors at either entrance are open at the same time.

(b) Write the corresponding Boolean expression.

# Answers

## Answers

*Chapter 2*
3    160 nA.

*Chapter 3*
1   (b) Each diode must withstand a maximum reverse voltage equal to the peak value of the a.c. input voltage.
2   (b) Each diode must withstand a maximum reverse voltage equal to twice the peak value of the a.c. input voltage.
4   (i) 80 Ω   (ii) 75 mA
5   (a) (i) 1·05 W   (ii) 58 mA
    (b) Breakdown in the 5 V device is likely to be due to the Zener effect. The p.d. across this device will therefore tend to decrease slightly with increased temperature. Breakdown in the 18 V device is likely to be due to the avalanche effect and as such the p.d. across it will tend to increase slighlty with increase in temperature. Using the two devices in series can therefore have the advantage of reducing the variation in the output voltage with temperature.

*Chapter 4*
2   2·5 kΩ
3   (i) 134   (ii) 60 kΩ
4   (i) 0·976   (ii) 0·998

*Chapter 5*
1   2·8 kΩ (2·7 kΩ), 525 kΩ (560 kΩ)
2   $Q_1$, $V_{CE} = 5·1$ V, $I_C = 2·1$ mA; $Q_2$, $V_{CE} = 8·2$ V, $I_C = 1·2$ mA
    Clipping will first occur on positive excursions of the output signal. The maximum voltage to which the collector can rise is 12 V. Hence, the maximum output signal is $(12 - 8·2) \times 2 = 7·6$ V p–p.
3   Assume $V_{R_3} = \dfrac{V_{CC}}{10} = 1$ V
    $R_3 = 670$ Ω (680 Ω); $R_2 = 10·7$ kΩ (10 kΩ)
    $R_1 = 50·9$ kΩ (47 kΩ); $R_L = 3$ kΩ (2·7 kΩ)

4 $A_v = -180$
5 $\beta$ deduced from the curves at $Q_1$ is approximately 100
 $A_v = -165$ (calculated)
 $A_v = -155$ (graphical)

*Chapter 6*
3 (a) 47·8 kHz  (b) 50·3 kHz
4 (a) 83·5 kHz  (b) 79·6 kHz

*Chapter 7*
1  0·56 ms
2  $t = 1$ ms, 2 ms, 3 ms
 0·7 ms (or 0·7 $\tau$). Note, this follows from the result deduced from Fig. 7.14 since 75 V is half of the applied voltage.
3  From growth curve for question 2, putting $v_c = 50$ V, $t = 0.4$ ms. Assuming the time for which $S_1$ is closed is negligible, $f = \dfrac{1}{t} = 250$ Hz
4  $\tau = 0.5$ ms hence output waveform as in Fig. 7.7.
5  $\tau = 1$ ms hence output waveform as in Fig. 7.6 except the signal has a peak-to-peak value of 10 V.
6  As Fig. 7.10 with the anode and cathode connections of $D_1$ interchanged. To get the tips of the negative pulses only, return $R_3$ to a negative supply.
7  2 nF
8  $C_1 = 2$ nF, $C_2 = 6$ nF

*Chapter 8*
2  $g_m = 2.1$ ms
3  $r_a = 22$ k$\Omega$ (interpolation necessary); $\mu = 48$
4  $v_o = 31$ V; $A_v = -31$

*Chapter 9*
2  (a) Increased horizontal deflection sensitivity.
 (b) Increased vertical deflection sensitivity.
 (c) Electrons in beam have higher velocity and so the horizontal and vertical deflection sensitivities are reduced.
 (d) Beam current increased and hence brightness increased.
4  The angle of deflection is unchanged but, since the coils are now further from the screen, the horizontal and vertical deflections are increased. If the coils are moved too far, the electron beam may strike the tube neck.
5  The c.r.o. trace will be deflected vertically by an amount proportional to the magnitude of the direct voltage. $Y$ shift may be achieved by this means.

*Chapter 10*
1  Figure 10.21 is equivalent to the simple machine control example

# Answers

in section 10.7, where the lamp $F$ represents the device to which the start signal is applied (i.e. logic circuit and truth table as in Fig. 10.14(a) and Fig. 10.14(b)).

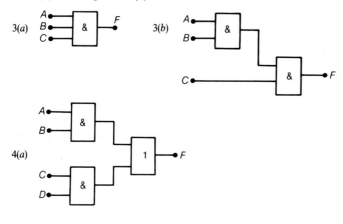

4(b)  $F = A.B + C.D$

# Index

Amplification factor, 122
Amplifier
  common cathode, 120
  common emitter, 59
AND gate, 145, 155
Anode
  c.r.t., 131
  diode, 118
  semiconductor, 19
  triode, 119
Astable multivibrator, 109
Atom
  energy bands, 2
  shells, 1
Avalanche breakdown, 20

Barrier potential, 12
Base, transistor, 39
Biasing
  transistor, 62, 74, 76
  valve, 126
Binary code, 147
Bipolar transistor—*see* transistor
Bridge rectifier, 24

Cathode
  c.r.t., 131
  diode, 118
  semiconductor, 19
  triode, 119
Cathode ray oscilloscope, 139
cathode ray tube, 130
charge curve, 96
charge drift, 3

clipper, 103
collector, 39
colpitts oscillator, 91
Control grid, 119
Common base
  a.c. input resistance, 45
  a.c. output resistance, 49
  current gain $\alpha$, 48
  input characteristic, 44
  output characteristic, 46
  relationship $\alpha$ and $\beta$, 56
  transfer characteristic, 49
Common emitter
  a.c. input resistance, 52
  a.c. output resistance, 56
  amplifier, 59, 76
  current gain, 65, 68, 70
  current gain, $\beta$, 54
  input characteristic, 51
  output characteristic, 53
  transfer characteristic, 56
Conductor, 3
Conduction band, 2
Covalent bond, 6
*C-R* network, 94
Crossover point, 134

D.C. power supply circuit, 32, 37
Deflection
  electrostatic, 133, 136
  electromagnetic, 137
Deflection coils, 139
Deflection plates, 136
Depletion layer, 12
Differentiator, 100
Diode
  semiconductor, 18
  valve, 115
  voltage reference (*VRD*), 20
  zener, 20
Diode gates, 155

Electron
  energy levels, 1
  gun, 131
  -hole pairs, 5

# Index

lens, 132
valence, 2
Electrostatic
deflection, 133, 136
focusing, 134
Electromagnetic deflection, 137
Emission
photo, 115
secondary, 115
thermionic, 114
Emitter, transistor, 39
Extrinsic conduction, 9

Final anode, c.r.t., 131
Focus anode, c.r.t., 135
Filament, 115

Gain
current, 65, 68, 70
power, 71
voltage, 66, 69, 125
Gates—*see* logic gates
Germanium, 4, 6

Hartley oscillator, 89
Heater, valve, 119
Holes, 4

Indirect heating, 118
Inductor-capacitor filter, 31
Insulator, 3
Integrator, 99
Intrinsic conduction, 4

*L.C.* oscillators, 81
Load line
transistor, 63
valve, 121
Logic gates
AND, 148, 155
INCLUSIVE-OR, 150
OR, 149, 156
NOT, 151

Majority carriers, 8, 9, 41
Minority carriers, 8, 9, 40

Monostable, 108
Mutual conductance, 122

$N$-type semiconductor, 6
Nucleus, 1

Operating point, 66
Oscillation
  damped, 83
  natural, 81
  sustained, 84
Oscillators
  Colpitts, 91
  Hartley, 89
  *L.C.*, 81
  tuned collector, 84

Peak inverse voltage (p.i.r.), 29
*P-N* junction
  barrier potential, 12
  *I-V* characteristic, 16
  leakage current, 17
  reverse breakdown, 19
Phosphor screen, 131
Positive feedback, 88

Quiescent operating point, 60

Ramp generator, 102
Recombination, electron-hole, 5, 12
Rectification
  full-wave, 23
  half-wave, 22
Reservoir capacitor, 27
Ripple, 27

Semiconductor, 3
  diode, 18
  $n$-type, 6
  $p$-$n$ junction, 11
  $p$-type, 30
Smoothing filter, 30
Static characteristics
  diode, 117
  transistor, 44, 46, 51, 53
  triode, 121, 123
Surge current, 30

# Index

Thermionic emission, 114
Timebase
　circuit, 112
　c.r.o., 140
Transfer conductance, 122
Transistor, bipolar, 39
　*see also* common emitter and common base
Transfer conductance, 122
Triode, 119
Triggering
　c.r.o., 142
　monostable, 108
Tuned collector oscillator, 84
Truth table, 145
Two-state systems, 146

Valence band, 2
Voltage reference diode, 20, 32
Voltage stabiliser circuit, 32

Working point, transistor, 60

$X$ plates, 141
$X$ amplifier, 141

$Y$ plates, 141
$Y$ amplifier, 141

Zener breakdown, 20
$Z$ modulation, 143